A Guide to IT Contracting

A Guide to IT Contracting

Checklists, Tools, and Techniques

Second Edition

Michael R. Overly

AN AUERBACH BOOK

DISCLAIMER: Information technology and contracting law changes frequently and rapidly. It is also subject to differing interpretations. It is up to the reader to review the current state of the law with a qualified attorney and other professionals before relying on it. Neither the authors nor the publisher makes any guarantees or warranties regarding the outcome of the uses to which the materials in this book are applied. This book is sold with the understanding that the authors and publisher are not engaged in rendering legal or professional services to the reader.

Second Edition published 2021
by CRC Press
6000 Broken Sound Parkway NW, Suite 300, Boca Raton, FL 33487-2742

and by CRC Press
2 Park Square, Milton Park, Abingdon, Oxon, OX14 4RN

First Edition published by CRC Press 2013

CRC Press is an imprint of Taylor & Francis Group, LLC

ISBN: 978-0-367-48902-1 (hbk)
ISBN: 978-0-367-76725-9 (pbk)
ISBN: 978-1-003-16646-7 (ebk)

Typeset in Garamond
by codeMantra

For Emma, the light of our lives these many years.

—Michael R. Overly

Contents

Acknowledgments

The author wishes to thank his colleagues, Chanley Howell and Steve Millendorf, for their patience, support, and many hours of work on this project.

About the Author

Michael R. Overly is a partner in the Information Technology & Outsourcing Practice Group in Foley & Lardner's Los Angeles office. As an attorney and former electrical engineer, his practice focuses on counseling clients regarding technology licensing, intellectual property development, information security, and electronic commerce. Michael is one of the few practicing lawyers who has satisfied the rigorous requirements necessary to obtain the Certified Information Systems Auditor (CISA), Certified Information Systems Security Professional (CISSP), Information Systems Security Management Professional (ISSMP), Certified in Risk and Information Systems Controls (CRISC), and Certified Information Privacy Professional (CIPP) certifications. He is a member of the Computer Security Institute and the Information Systems Security Association. Michael is a frequent writer and speaker in many areas, including negotiating and drafting technology transactions and the legal issues of technology in the workplace, e-mail, and electronic evidence. He has written numerous articles and books on these subjects and is a frequent commentator in the national press (e.g., the *New York Times*, *Chicago Tribune*, *Los Angeles Times*, *Wall Street Journal*, ABCNEWS.com, CNN, and MSNBC). In addition to conducting training seminars in the United States, Norway, Japan, and Malaysia, Michael has testified before the U.S. Congress regarding online issues. Among others, he is the author of the best-selling *e-policy: How to Develop Computer, E-mail, and Internet Guidelines to Protect Your Company and Its Assets* (AMACOM 1998), *Overly on Electronic Evidence* (West Publishing 2002), *The Open Source Handbook* (Pike & Fischer 2003), *Document Retention in the Electronic Workplace* (Pike & Fischer 2001), and *Licensing Line-by-Line* (Aspatore Press 2004).

About the Author

Preface to the First Edition

Introduction

Information technology (IT) is critical to the operation of every business. IT drives enterprise efficiency by breaking down communication barriers both internally among employees, management, and directors and externally between the business and its customers, advisors, and other contributors. Unfortunately, in our cumulative decades of practice in the area of IT and outsourcing law, we have found that many businesses fail to identify and adequately address the issues that arise in their IT contracts. Together, we have reviewed, drafted, and negotiated thousands of IT contracts for practically every type of product, software, hardware, and service. Time and time again, we have found businesses fail to address the essential elements of their contracts and, ultimately, fail to adequately protect their organizations, placing their assets and data at risk. This can expose the company to unnecessary liability and lead to uncontrolled costs. The net result? Companies assume far greater risk and liability associated with these transactions, and end up spending too much time and money on these transactions, than is necessary.

Even the most sophisticated businesses with the most sophisticated IT infrastructure and an army of lawyers in their in-house legal departments are frequently unfamiliar with all the key issues that may arise in contracting of this kind and are, therefore, not equipped with the tools or knowledge to make informed and strategic decisions when reviewing, drafting, and negotiating these contracts. What's worse, companies are accustomed to using outside corporate lawyers who might be highly skilled transactional and securities lawyers, but who aren't familiar with the nuances of IT or how to properly structure an IT deal to protect the client and its assets.

IT is likely not your company's core competency or your company's focus. Nonetheless, it likely drives your business. Your board may cringe when the CIO asks for millions of dollars' worth of upgrades to IT infrastructure or approval for that next big software or outsourcing project. IT spending might not be your company's largest expense, but it is probably among the top two or three. While your company may not focus on IT and might constantly question the millions of dollars' worth of equipment, software, and professional services that comprise your IT infrastructure, it's a sure bet that the company wouldn't survive long without it. Think about what your day would be like today if your e-mail system was down for even an hour or two, let alone several days. What would happen if your website suddenly crashed and the company could no longer take orders? How much would productivity suffer if your document management system was suddenly inaccessible, or if the data stored in the system disappeared? What would the reaction be if your payroll system went down and people didn't get paid? How much lost productivity would result if your customer relationship management (CRM) system crashed and your sales force couldn't access information it needed to contact clients and sell products? What would the result be if your company's data security system failed and customer and employee data was suddenly accessible outside of the company?

When IT fails, businesses may lose revenue, suffer harm to their reputation, compromise the security of and lose data, face regulatory sanctions and liability, lose competitive advantage, and be exposed to lawsuits from their business partners, customers, investors, employees, and shareholders. Failure of IT can literally mean that the "lights go out" for a business. The fact that IT is a (if not the) most critical component of your company, without which your company may suffer considerable harm, is the reason we wrote this book.

This book is designed as a Rosetta Stone of sorts for IT contracting. It presents a distillation of the most critical business and legal lessons learned from decades of drafting and negotiating virtually every type of IT-related agreement. In a single volume, you can quickly access information on almost every type of technology agreement and immediately understand what the agreement is intended to do, the critical business and legal issues that must be addressed, tips and best practices for negotiation of key terms of the agreement, and the most common pitfalls typically encountered. The information is presented in a highly accessible handbook form, intended to provide you with immediate tools to more intelligently address issues in technology contracting and how to negotiate those types of agreements.

Each chapter focuses on a particular type of IT agreement (e.g., software license agreement, hosting agreement, professional services agreement, development agreement, cloud agreement). A checklist of essential terms that are commonly included in the particular agreement is followed by a brief summary of what the agreement is intended to do and how it is used. The summary is followed by a thorough review of key legal and business issues that are commonly included and addressed in the particular agreement, with common pitfalls clearly highlighted. Each chapter concludes with a summary of key issues.

A glossary is included at the end of the book to provide ready access to common terms and acronyms encountered in IT contracts.

A CD is included with copies of the summary checklists provided throughout this book.

Our goal is to provide you with a resource you will use for years to come. Whenever you are presented with a new IT contract, you need only to open the relevant chapter to quickly acclimate yourself to the unique issues presented by that type of agreement. If there are additional types of agreements that you would like to see included in subsequent editions of the book, please let us know.

The importance of IT to a company cannot be understated. Simply put, without IT, businesses will cease to function. Unfortunately, too many factors, such as poorly drafted vendor forms, corporate time pressure to enter into agreements quickly (and get programs up and running quickly) to drive cost savings, lack of expertise with the types of contracts used in IT contracting, and relying on service providers (lawyers, contractors) who don't understand the essential (and the nonessential) elements of contracting for IT products and services, can lead companies down a path to bad contracts that subject the company to unnecessary expense and risk.

We have seen time and time again companies that choose vendors to achieve cost savings, rush through the contracting process, and are left with little in the way of remedies when the vendor can't perform as promised. We are frequently left on the other end of the phone with the customer asking, "What can we do?" Poorly drafted IT contracts have led to vendors repudiating transactions; lawsuits costing both sides hundreds of thousands of dollars, lost productivity, months and years of delay; data disasters involving government regulators; customers being left without a product, without their data, and without any remedies; and the all too frequent data security breach in which the vendor has no responsibility and the customer has no remedy.

Mistakes can be avoided, risk can be averted, and better contracts can be drafted if you have the information necessary to make informed decisions about how much risk your company is willing to take and what contractual terms in your particular agreement are commonly used to minimize that risk. Our hope is that this book sets you down that path.

How to Use This Book

As with most contracts, when drafting and negotiating an IT contract, it is always best to use a methodical approach. Following the recommended steps below will help you to get the most from this book.

Know what you are getting before you start and make sure you have the right agreement. Before you put pen on paper, be sure you understand what type of agreement you are reviewing. Don't rely solely on the title of the agreement or what your business people tell you the agreement is. Too many times we have seen cloud agreements called end-user license agreements and IT professional services agreements called software license agreements. It is frequently the case that you may be told you are contracting for a hosted application, but the agreement you actually get is for a locally installed application. Read the contract.

Ask your business partners and other stakeholders what they are buying or licensing. Decide, is this a service agreement? An agreement under which something is being licensed? Is the vendor providing some form of hosted or cloud-based service? Determine whether you have the right agreement before you start drafting and negotiating. While this first step seems trivial, the answer is frequently unclear and requires investigation. In some cases, the vendor may be providing a combination of things (e.g., a software license and associated professional services to implement the software, a cloud service that also has a locally installed component). Make sure you understand what you are reviewing, what the product or service is, and whether you are reviewing the correct agreement. If you aren't certain, ask someone and keep asking until you get a clear answer. It is always disruptive to the contracting and business process when we are asked to review a contract, spend time (and our client's money) reviewing it, begin negotiations, and only then have a business person say, "wait, we aren't running this on our servers, we are getting a cloud-based application and need to be reviewing that agreement." Then, we must start the process all over again. This represents a loss of time, money, and efficiency, not to mention the frustration on both sides of the transaction.

Find your agreement in this book. After you have determined the type of agreement you have, identify the relevant chapters in this book that discuss that type of agreement. Read through the quick checklist to survey the landscape and assess what might be missing from the agreement you are reviewing. Then read the introductory material in those chapters to better understand the intent behind the agreement you're using. Use the checklist and content when you are drafting to ensure that you have covered, or at least considered, everything that should be contained in the relevant type of agreement.

Understand the business requirements. Identify the relevant business requirements (e.g., what are the anticipated fees, how critical is the engagement, will sensitive data be used or accessed by the application or as part of the professional services, are there any unique specifications, what is the term of the agreement, how many of the supplier's people are needed to provide the services and are there specific people that are required, when will the supplier be required to begin and finish the work, will services be provided remotely or at your company's place of business, and who is paying for travel and related costs). Involve all relevant stakeholders (e.g., business, information security, compliance, legal, HR) in the business in this process.

Review the detailed best practices and key issues in the relevant chapters of this book. Compare the relevant sections of the particular chapter with the contract you are reviewing. Are there things missing? Would additional language help mitigate the risk your company faces by agreeing to the existing language in the contract? What changes to the contract are essential to your business? In addition to those considerations, add any other unique requirements of your business. For example, you may be a healthcare provider considering an engagement that will require the vendor to have access to protected health information. You will want to consider using (and may be required to use), at minimum, a Business Associate

Agreement to supplement the underlying contract. Or you may have certain standards with respect to confidentiality, data security, personnel requirements, and the like that you include in all of your agreements. Be sure to evaluate whether those standard company provisions must be included in the contract you are reviewing.

Review the draft contract with the foregoing in mind. Identify relevant sections for revision and revise accordingly. Discuss provisions with appropriate stakeholders within your organization (e.g., risk management for the insurance provision, IT/security for the relevant technical and security provisions).

Make a list of terms you are not familiar with. Identify any technical or contracting terms that are unfamiliar and review their definitions in the glossary at the end of this book. If the term is not in the glossary, ask your IT people what the term means and ensure that you understand it.

Use the checklists to ensure all relevant issues have been identified. It is not likely that you will able to include all of the terms identified in the checklists in each agreement of that particular type. However, it is important to think through each term and how it relates to your business and consider each term and whether it should be included in the contract that you are reviewing. If you are unable to incorporate a particular provision into the contract you are negotiating, be prepared to articulate why and understand the risk to the company of not including those provisions. Present these to the stakeholders to ensure they understand the risk and can make the appropriate business decisions with respect to whether to move forward.

Preface to the Second Edition

Since the first edition of this book, there have been several alarming trends in the technology contracting industry. Several of these trends are described below. As a technology customer, you need to be ready to explore alternate solutions from other vendors, conduct simultaneous negotiations with other vendors, and, generally, ensure prospective vendors understand they can "lose the deal" if they refuse to act reasonably. Even then, in some cases, very material risks must be accepted to move forward with a vendor. Those risks should be identified to senior management as early as possible in the process so that appropriate decisions can be made.

The Dawn of the "As-Is" Technology Product. Over the last two to three years, we have seen a very steady movement among vendors to ever more completely avoid responsibility for their actions and products. This has culminated in some vendors, even in deals involving several hundred thousand dollars in fees, offering technology solutions completely as-is. That is, no warranties at all. The technology need never work, need never operate properly, and need never be available for use. In exchange, the customer is asked to sign an irrevocable commitment to pay for the technology.

The Ever-Changing Product. It is not at all uncommon today to find technology products where the functionality, performance, security, etc., of the product are described through a referenced website, which can change at any time. The vendor can materially decrease the overall functionality of a product, change its security procedures for the worse, etc., and the customer cannot object or terminate the agreement.

Where in the World Is My Data? It is a frequent occurrence that vendors may use numerous offshore affiliates and subcontractors to perform their obligations, including hosting, manipulating, and storing the potentially highly sensitive information of the customer (e.g., sensitive intellectual property or regulated personally identifiable information). The security of those offshore facilities, management of their personnel, and intellectual property and privacy laws applicable in these other countries may fall far short of what is required to adequately protect the customer's information. In many cases, the vendor, itself, is unable to say at any given time where the customer's data resides and who has access to it. Some vendors proudly announce that they use "best in class encryption," but then refuse material liability if their handling of the encryption keys or construction of the encryption algorithm is flawed.

The foregoing makes technology contracting ever more difficult and highlights the need to follow the recommendations in this book to mitigate potential risk. It also makes clear the importance of maintaining negotiating leverage with potential vendors. All too often, vendors start negotiations knowing "they have the deal" and have no motivation to make reasonable concessions. Ensuring the vendor knows "they may lose the deal" is the best way to maintain leverage in negotiations.

Chapter 1

Collecting Basic Deal Information

CHECKLIST

Basic Principles

- ☐ Marshal basic information
- ☐ Value of proposed transaction?
- ☐ Term of agreement?
- ☐ Criticality of technology to business?
- ☐ Unique regulatory issues?
- ☐ Other foundational information?
- ☐ Circulate a "deal memo"
- ☐ Circulate a "term sheet"

Describe the Engagement

- ☐ What is the deal about?
- ☐ Business advantage from contract?
- ☐ Use nontechnical English

Useful Life

- ☐ Anticipated duration of contract
- ☐ Desired renewal terms
- ☐ Duration of services rendered

- ☐ License for years or perpetual?
- ☐ Renewal rights
- ☐ Costs for renewal

Expected Fees

- ☐ Compensation to vendor
- ☐ Breakdown of first-year fees
  - License
  - Professional services
  - Implementation
  - Customization
  - Hardware
  - Telecommunication
- ☐ If no fees, good faith estimate
- ☐ When to use customer's form

Performance

- ☐ Customer-facing application?
- ☐ Location for service performance?
- ☐ Offshore vendor?
- ☐ Vendor uses offshore partners/affiliates?
- ☐ Vendor uses subcontractors? If so, who?
- ☐ Location for vendor performance?
- ☐ Vendor provides hosting services?

Intellectual Property

- ☐ Will the customer want to own vendor-created IP?
- ☐ Vendor cannot share with competitors?
- ☐ Vendor cannot share with industry?
- ☐ Vendor has access to sensitive IP?

Personal Information

- ☐ Vendor access to personally identifiable information?
- ☐ What information is at risk?
- ☐ Financial account information?
- ☐ Health information?
- ☐ Social Security Numbers?
- ☐ Legal and regulatory requirements
- ☐ Transmission across international borders

Information Security

- □ Vendor access to sensitive customer data?
- □ Cloud computing–based service?
- □ Hosting service?
- □ Is vendor sole custodian of customer data?

Unique Issues

- □ Vendor's financial situation is suspect
- □ Vendor is subject of litigation
- □ Vendor had recent security breach
- □ Performance constraints
- □ Substantial regulatory/compliance issues

Overview

Before any proposed technology contract can be reviewed, certain basic information about the deal must be marshaled. This includes the value of the proposed transaction, the term of the agreement, the criticality of the technology to the business, how long to implement, unique regulatory issues (e.g., is sensitive personally identifiable data at risk?), and other foundational information. While this process may seem self-evident, it is common for businesses to rush forward in the review of a proposed technology contract without this critical information.

In our experience, moving forward without a clear understanding of the "deal" can result in misunderstandings with the vendor, failure to achieve an adequate and appropriate contract, delays in negotiations, and increased costs. For example, it would likely not be appropriate to require the same level of contractual protection in a $20,000 off-the-shelf license agreement as one would require in a $20 million custom software development deal for a critical client-facing application. Similarly, it would probably not be fruitful to propose extensive information security language in a contract that does not involve highly sensitive information. Finally, it would be all but impossible to impose one-off service level obligations in a small, noncritical Application Service Provider (ASP) deal, but those obligations may well be entirely appropriate in a large-scale transaction.

While the foregoing may seem obvious, we frequently see businesses proposing contract terms that are inappropriate for the contemplated engagement. In most cases, the problem arises from a failure to assess the transaction adequately from the outset. The reason for that failure is almost always a lack of clear foundational information about the transaction.

In this chapter, we identify key areas for which information should be obtained and understood before any review of draft contracts commences. By assessing this information, businesses can make more informed decisions about their proposed technology engagements and ensure their contracts are appropriate to those engagements. This list is not intended to be exhaustive. Other issues unique to a particular transaction or business should be added and, of course, the vendor should conduct appropriate due diligence.

Many businesses now require the foregoing information to be recorded in an internal "deal memo" and circulated to all relevant stakeholders (e.g., risk management, legal counsel, information security, and, of course, senior decision-makers). The deal memo is different from and should not be confused with a "term sheet," which is designed to summarize the business terms of the deal. The term sheet will be circulated between the vendor and the customer. The deal memo is intended to be a purely internal document to educate relevant company stakeholders regarding the transaction. An example deal memo is provided at the end of the chapter.

Key Considerations

- **Executive description of engagement.** Write a sentence or two describing in plain, nontechnical English what the deal is about, including a clear statement establishing the business advantage to be gained by entering into the contract. For example, "The license of a new expense tracking application designed to identify duplicate expenses more readily. Expected cost savings are projected to be $500,000 per year." This type of description helps all those involved in the process immediately understand the nature of the transaction.
- **Useful life.** Establish the anticipated duration of the contract, including desired renewal terms. In particular, if professional services will be rendered, what is the expected duration of those services (e.g., if software will be implemented, the duration of that implementation)? Will services take only a few weeks or will the services extend over many months? The longer the term of services, the greater the need for contractual protections relating to project management and cost control. Similarly, if a license is being granted, is the software being licensed for a term of years or perpetually? If for a term of years, the agreement should address renewal rights and the costs for renewal, including price protection for those renewals. Technology is constantly involving. In many cases, leading-edge products today are yesterday's news in just a few years. This is why many technology agreements generally have relatively short initial terms (e.g., 2–5 years). The customer needs the ability to move to the next "big thing" and not be tied to outdated technology.
- **Expected fees.** Describe the compensation due to the vendor over the life of the contract, including a breakdown of all first-year fees (e.g., license, professional services, implementation, customization, hardware, and telecommunications fees). If the fees cannot be completely defined at the outset, good faith

estimates should be provided. If estimates are not possible, evaluation of the deal is likely premature. Even a "ball-park" figure can greatly assist in determining the approach to the contract. As a general rule, the larger the engagement, the greater the leeway the vendor will have in entertaining proposed revisions to its contract or in considering the customer's form agreement. That is, while it may be completely inappropriate for a customer to attempt to use its own form agreement for a $10,000 off-the-shelf software license, it may be entirely appropriate to use the customer's form in larger transactions.

- **Understand the solution.** Make sure the vendor's solution is fully understood. Is it a software application that is locally installed on the customer's own systems? Is it a cloud solution residing on the vendor's own servers? Does the vendor have no technology infrastructure and its cloud solution is actually hosted using a third-party hosting facility? Is the vendor's solution a combination of locally installed software and cloud services? All too frequently, these basic questions are not answered before review of the proposed contract is commenced.

Performance

- Consider how critical this service or product is to the company.
- Is this a customer-facing application?
- Where will the services/hosting be performed?
- Is the vendor located offshore?
- Will the vendor use offshore partners or affiliates?
- Does the vendor require the use of subcontractors? If so, who are the intended subcontractors?
- Will the vendor be performing services onsite or at its own facilities?
- Will the vendor be providing hosting services of any kind (e.g., Cloud Computing, SaaS, ASP)?

Intellectual Property Issues

- Will the vendor render any development or customization services resulting in the creation of intellectual property that the customer will want to own?
  - If not owned by the customer, at minimum, should there be a period of time when the vendor cannot share the intellectual property with the customer's competitors or in the customer's industry? This type of "cooling off" provision will ensure the customer doesn't foot the bill for its competitors to gain an advantage over it.
- Will the vendor have access to any highly sensitive company intellectual property?

Personal Information Privacy and Security

- Will the vendor have access to or possession of sensitive, regulated personally identifiable information? If so, what information will be at risk?
  - Will the vendor have access to sensitive personal information, such as financial account information, health information or Social Security Numbers?
  - What are the applicable legal and regulatory requirements?
  - Will the data be transmitted across international borders?

Information Security

- Will the vendor have access to or store confidential, proprietary, or sensitive information and data of the customer? If so, what information will the vendor have access to or store?
- Is the vendor providing a cloud computing–based service or hosting services? This can raise the criticality level significantly depending on nature of information stored and controlled by the vendor.
- Will the customer retain a copy of backup data stored or hosted by the vendor, or will it rely on the vendor as the sole custodian?

Other Unique Issues

- The deal memo should also identify any other unique issues to the proposed transaction, including but not limited to:
  - Unique business risks (e.g., the vendor's financial wherewithal is suspect, the vendor has been the subject of recent litigation, the vendor has recently had a data security breach)
  - Unusual performance constraints (e.g., project must be completed within a defined period of time, the project is business critical, project is part of a larger technology initiative that relies on this component)
  - The engagement raises substantial regulatory or compliance issues.

Summary

By investing a minimal amount of time to marshal basic deal information at the outset of any new engagement, businesses can be far more prepared to negotiate their vendor agreements effectively. Understanding and assessing this information will ensure vendor agreements are appropriate to the relevant transaction and that negotiation time and expenses are minimized.

Example Deal Memo

Confidential/Company Proprietary

Vendor Name:

Address:

Primary Vendor Contact:

Executive Overview of Transaction:

Transaction/Technology Life Span:

Projected Fees for Life of Engagement:

Performance Requirements:

Intellectual Property Ownership:

Personal Information Privacy and Security:

Information Security:

Other Unique Issues:

Prepared by: ______________________________

Title: ______________________________

Date: ______________________________

Chapter 2

Software License Agreements

CHECKLIST

Ask Yourself:

- ☐ Business purpose and goals
- ☐ Criticality
- ☐ Fees and costs
- ☐ Implementation time

Terms to Include

- ☐ License and restrictions
- ☐ Acceptance testing
- ☐ Third-party software
- ☐ Fees
- ☐ Warranties
- ☐ Indemnification
- ☐ Limitation of liability
- ☐ Specifications
- ☐ Confidentiality and security
- ☐ Maintenance and support
- ☐ Announcements and publicity

- □ Term and termination
- □ Additional terms:
  - Force majeure
  - Prohibition on right to assign

Introduction

While every software license agreement does not warrant a customized approach, it is important that software license agreements contain the essential terms that address relevant business issues. It is often the case that software licensing transactions are driven by a vendor's software license agreement. Understanding that, in many cases, it makes sense to start with a vendor form of license agreement, this chapter discusses the language that your company should expect to see in its software license agreements. Short vendor forms are used by vendors in transactions that range from the simple to the complex and may require a small financial commitment or several million dollars. When reviewing, drafting, and negotiating software license agreements, it is essential that your company understand each provision of the vendor agreement, how it operates, what changes should be made to ensure that the provisions meet your company's business requirements, and what additional provisions your company should propose adding to the software license agreement so that the result is an agreement that is balanced and meets your company's needs.

Four Critical Questions

Before drafting or reviewing any software license agreement, your company should ask four critical questions. The answers to these questions will, in large part, dictate the type of provisions that you draft and the depth of revisions that you will propose to the vendor's form of license agreement. Recall similar questions, but in a general context, were described in Chapter 1. These questions are targeted at software license agreements in particular.

- **Business purpose and goals.** What is the business reason for licensing this application? This should include a fairly robust discussion that the business has for implementing the application.
- **Criticality.** How critical is the licensed application to the organization? By asking this question, your company is determining whether the application supports a critical business function or a minor business function. Your company should ask itself—what are the implications to the business if the application doesn't work?

- **Fees and costs.** What is your company going to pay for the application—including with respect to the initial license, support, implementation fees, professional services fees, maintenance, etc.
- **Implementation time.** How long will it take to implement the software? The longer the implementation period, the greater the likelihood for cost overruns, delays, and other problems. Where implementation is an issue, the parties will want to develop a specific work plan and schedule prior to execution of the software license agreement and, at a minimum, determine the commencement date of the implementation, the required completion date, and all of the dates associated with the key deliverables of the implementation. Where a complete work plan is not determined prior to agreement execution, ensure that the agreement includes a requirement that the parties work together to finalize the work plan within a specified number of days after the agreement becomes effective.

License and Restrictions

Each license agreement should contain a license to the software that provides all authorized or possible users access to the software.

- Ensure that your company can assign the license if it requires this right.
- Ensure that successor entities to your company can use the application for a period of time in order to ensure a smooth transition. In this case, the vendor will likely require that your company remain responsible for its obligations, including payment obligations, for the period of time the successor entity is permitted to use the software.
- Ensure that, if applicable, your company is permitted to outsource the operation, hosting and use of the application and that such right is without charge.

It is generally preferable to avoid limiting the use of software for "internal purposes" and instead broadening the license grant to provide your company the ability to use it for its "business purposes." In particular, think about your customers, business partners, your contractors, your suppliers, etc. Do they need to use the software? If so, ensure you negotiate those rights into the agreement.

- The "licensed software" should be defined to include all of the software that your company is licensing. Keep in mind that this likely includes any updates, enhancements, modifications, versions, and customizations to the software.
- The application vendor will likely want to include broad restrictions in the license grant (i.e., things that your company can't do with the software). Be sure to read these carefully and tailor them to ensure that the restrictions do

not limit your company's ability to use the software as required to meet your company's business objectives. These may include, for example, that your company cannot operate the software in a "service bureau" environment or for the benefit of unrelated third parties, remove any proprietary notices from the software, or disassemble, decompile, or reverse engineer the software.

- Consider whether your company will be charged for software updates. If your company is not going to be charged for updates, whether permanently or for a specified period of time, this should be stated clearly in the software license agreement.
- It is important to determine what the term of the license will be. In many circumstances, a perpetual license is required by the customer and offered by the vendor. In other circumstances, the parties will agree to a term license, limited to a specified number of months or, more commonly, years. If the parties agree to a term license, it is necessary to consider how the agreement will be renewed—i.e., will there be automatic renewals or will the parties be required to agree to each renewal? There are only rare circumstances where the vendor should be involved in the renewal discussion or have the ability to restrict the customer's renewal of a term license.
- Consider whether the license extends to only the object code or the source code as well. In many circumstances, it is appropriate to require that the vendor establish a source code escrow, which requires the release of the source code to the customer upon the occurrence of certain release conditions (i.e., in situations in which the license doesn't already include and the vendor hasn't already delivered the source code).

Acceptance Testing

When delivered, your company may want the right to evaluate, test, and accept the applications and any deliverables that the vendor delivers to your company under the software license agreement.

- It is rarely acceptable to have an automatic, or "deemed," acceptance of software (i.e., acceptance by your company purely as a function of the passage of time).
- Acceptance and payment terms are commonly tied together. Consider making some payment terms contingent upon your company accepting the application.
- Similarly, it is a good practice to hold back a certain portion of all of the fees (e.g., 15%) until final acceptance (e.g., acceptance of all of the software that is to be provided by the vendor).
- Vendors will commonly require and it is generally acceptable to agree to a period of time during which acceptance can occur (e.g., within thirty days

after delivery of the software to the customer). Beware of vendors that insist on adding language that states if the customer fails to reject within the defined period of time, the software will be "deemed accepted." Ordinarily, we try to avoid such language. If the vendor will not agree, make clear that deemed acceptance does not result in the waiver of any other right or obligation elsewhere in the agreement. For example, deemed acceptance will not result in a waiver of any warranties.

- A process for acceptance testing should be included in the acceptance testing provision of the software license agreement (e.g., the required time periods, requirements of the vendor with respect to correcting software or deliverables that are not accepted, number of times the acceptance testing process will repeat, customer remedies in the event the software is never accepted). A separate exhibit or statement of work would typically contain the specific acceptance testing criteria (i.e., what are the elements that must be achieved in order for the software to be accepted?).

Third-Party Software

Third-party software is commonly included by vendors in their software. Sometimes such third-party software is integrated into the application that your company is licensing. Other times, the vendor will package additional applications into a suite of products that you are licensing. Often it is a combination of the two.

- If your company must accept third-party terms and conditions with respect to the third-party software, your company must have an opportunity to review the applicable terms and conditions, and all such terms and conditions should be attached as an exhibit to the software license agreement. In particular, beware of flow through third-party terms that contain broad indemnities from the customer or audit rights allowing the third party unbridled access to customer systems and records.
- The vendor should warrant that it has the right to license the third-party software that is licensed under the software license agreement and that such third-party software does not infringe the intellectual property rights of any other party.
- The vendor will commonly agree to support the third-party software to the same extent that it supports the vendor's software that your company is licensing.
- Any open-source software that is included in the vendor's software should be specifically identified in an exhibit to the software license agreement, and all of the foregoing terms should apply to both third-party software and open-source software (which should be included in the definition of third-party software).

Fees

While the actual fees to be paid by the company to the vendor are commonly set forth in an exhibit to the software license agreement, the agreement terms and conditions typically contain fee and payment terms. As a starting point, your company should choose an appropriate metric for pricing and use restrictions:

- An enterprise license is the broadest, and generally the most expensive, license because your company is granted use of the software on an enterprise-wide basis.
- A named user license requires that specific people be "named" in order to have access to the software and the license is limited to a specified number of named users.
- A concurrent user license is a license grant to a specific number of users that can simultaneously access the software at any given point in time.
- A per-seat license grants the rights to use the software to specific workstations.
- A per-device license requires that the software only be used on specific hardware devices.

After your company has determined the appropriate metric for pricing, additional pricing and fees terms should be considered.

- Payment terms (for example, fees are due and payable by your company to the vendor within thirty days after receiving a valid and undisputed invoice). Note that the time your company has to pay an invoice should run from the date that the invoice is received by your company and not the date that is on the invoice or the date that the invoice is generated.
- Consider adding an "all fees" clause that states specifically that, unless stated in the agreement or in the fees exhibit, there are no other fees to be paid by your company to the vendor. This is a critical way to prevent unanticipated costs.
- The responsibilities of each of the parties with respect to taxes should be clearly delineated in the software license agreement. Consider also adding a clause that requires that both of the parties work together to minimize the tax liability that may arise, to the extent legally permissible.
- A process with respect to disputing charges and fees should be included in the software license agreement.
- Consider adding a clause to the software license agreement that makes it clear that any payment of fees by the customer to the vendor does not, under any circumstances, imply customer's acceptance of the services or any deliverables to be provided by the vendor under the agreement.

Warranties

When using vendor forms as a starting point for software license agreements, you will quickly notice that vendors typically disclaim all liability with respect to the licensed software and all warranties with respect to the licensed software. As we discussed earlier, the nature of the warranties that you seek, and the number of them, will depend on your company's answer to the questions posed early in this chapter, particularly with respect to how critical the software that you are licensing is to your company. In this section, we have described some of the more common warranties in software license agreements.

- The software will perform in accordance with the "specifications." Typically there is a period of time during which this warranty will apply (e.g., twelve months after acceptance of the software by the customer). Beware the trend in the industry is to shorten the duration of warranties, with some warranties only lasting thirty (30) days.
- The vendor has the authority to enter into the software license agreement and to grant the rights (particularly the license rights) that are contained in the software license agreement.
- The vendor's performance under the software license agreement does not violate any other agreements that it is a party to or any laws.
- There is no pending or threatened litigation that would have a material impact on the vendor's ability to perform its obligations or grant the rights that it grants under the software license agreement.
- The software will conform in all material respects to the documentation, specifications, and any other applicable materials.
- Your company's permitted use of the software will not violate the intellectual property rights of any third party.
- The software does not and will not contain any destructive mechanisms such as viruses, time bombs, worms, trap doors, and the like.
- Any services that are provided by the vendor will be provided in a professional and workmanlike manner.
- The software will comply with all federal, state, and local laws.
- The vendor will promptly correct and repair any deficiencies in the software.

Many software license agreements also contain warranties with respect to the vendor supporting and providing updates to the software for a predetermined period of time (e.g., ten years from the effective date of the software license agreement). This can be used as a mechanism to prevent the software that your company is using from becoming obsolete.

As noted in the preface to this second edition, there is a small, but growing trend among vendors to offer their solutions complete as is, without warranties of any kind. That approach may be entirely acceptable in the context of a $5,000 off-the-shelf software purchase for a noncritical application, but unworkable for larger, more important transactions.

Indemnification

Many, if not most, information technology contracts contain provisions that require one party (the indemnitor) to indemnify the other party (the indemnitee) for third-party claims arising out of certain events. Most commonly, software license agreements contain indemnification obligations of the vendor with respect to third-party claims against the customer that the software infringes the intellectual property rights of the third party.

- The intellectual property infringement clause should extend to claims arising out of a claim that the software infringes any third-party intellectual property including any trademark, trade secret, copyright, patent, and any other intellectual property or proprietary rights.
- Include the vendor's obligation to indemnify, defend, and hold the customer harmless from all liability including any damages and expenses incurred, arising out of or relating to the third-party claim.
- The vendor will typically seek and it is common in these provisions to include a clause that requires the customer to promptly notify the vendor of the claims and to cooperate in the defense or settlement of the claim.
- Include a provision that provides a remedy to the customer in the event of such a claim or the vendor determines that the software is likely to become the subject of a third-party intellectual property claim. These types of remedies commonly include the obligation of the vendor to replace the infringing or potentially infringing software with noninfringing software that contains the same functionality as the replaced software or the obligation of the vendor to promptly procure the right for the customer to continue using the software that is the subject matter of the claim or the potential claim. If neither one of the remedies is available, the vendor will typically require that the customer return the software, the agreement would then terminate, and the vendor should be required to return any license fees paid by the customer to the vendor for the infringing software.

Depending on the nature of the software license agreement and the software and services to be provided by the vendor, there are other indemnities that are commonly included in these types of transactions. These include indemnities with respect to third-party claims arising out of the vendor's failure to comply with applicable laws, property damages, and the vendor's negligence or misconduct.

Limitation of Liability

Perhaps the most discussed and negotiated area of any information technology agreement is the limitation of liability. This is usually an area where the parties begin discussions from very different vantage points, and if using vendor's forms as a starting point for your software license transactions, you will quickly notice that most vendor forms include a cap in the vendor's liability, but no protection for the customer.

There are commonly two types of damages that are recoverable in a software licensing transaction—consequential damages and direct damages. It is common for the parties to disclaim all liability, with some exceptions, for consequential damages that may arise, while the ability to recover direct damages is commonly capped at a specified amount of some multiple of the fees paid by the customer to the vendor under the software license agreement.

- Ensure that the limitation of liability is applicable to both parties. As mentioned, if you are starting with a vendor form of software license agreement, you will quickly notice that most vendor forms contain a limitation of liability that is only applicable to the vendor.
- When thinking about the cap on direct damages, think about the damages that could arise. Your company will want to choose a dollar amount, or a multiple of fees, that appropriately allocates responsibility for damages your company suffers to the vendor.
- Consider whether instead of making the one-sided limitation of liability mutual, your company proposes deleting the limitation of liability all together. In many instances, particularly in the software license agreement context, this can be an advantageous position.
- Ensure that certain things are "carved out" of the limitation of liability—both the exclusion of consequential damages and the cap on direct damages. This should, at a minimum, include damages that arise out of the vendor's breach of its confidentiality obligations, claims for which the vendor is insured, and damages payable pursuant to and expenses incurred as a result of the vendor's obligations to indemnify the customer under the agreement. In many instances, it is also appropriate to include in this list damages that result from property damage or injury to employees.

Specifications

As mentioned previously, software vendors typically tie all warranties, representations, and vendor performance with respect to the software to the "specifications" or "documentation." It is important to understand that the vendor's "specifications" and "documentation" are completely within the vendor's control and are subject to change at any time, without notice to the customer. What's more, many companies don't take the time before the software license agreement is signed to read the specifications or documentation to ensure that they understand the content. Control over the documentation and specifications gives the vendor an enormous amount of control over what they are contractually responsible for providing to you. They can change features and functionality at any moment and could ultimately alter the value of the software to your company. In order to avoid these issues, consider incorporating the following terms into your company's software license agreements:

- If there are specific features or functionality that must be contained in the software, these should be identified in the body of the agreement, or perhaps more appropriately, in an exhibit to the agreement.
- In many cases, it will be appropriate to include the vendor's specifications and documentation as of the effective date of the software license agreement as an exhibit to the agreement. The vendor might require the ability to modify these specifications and documentation over time. If that is required, ensure that the agreement states clearly that the specifications and documentation included in the agreement provide a baseline with respect to the features and functionality of the software and that any changes to the specifications and documentation will not diminish the features and functionality contained in the software as of the effective date of the software license agreement.
- Consider adding a broad term "software specifications" and define it to include all specifications, documentation, materials, functionality, request-for-proposal (RFP) responses, etc. If you do this, ensure that all of the vendor's warranty and other obligations are tied to the newly defined software specifications.

Confidentiality and Security

Like the other provisions that we have discussed in this chapter, the confidentiality clauses in vendor form software license agreements are typically drafted very narrowly and are one-sided in favor of the vendor.

- Ensure that the confidentiality clause is mutual and protects the company's information as well as the vendor's information.
- Read the definition of confidentiality and ensure that it is drafted to include all of the types of information that your company may be sharing with the

vendor or that the vendor will have access to. Generally speaking, a broad definition of confidential information is favored.
- While it is common in confidentiality provision in software license agreements to exclude certain types of information (e.g., information that was already known by the other party or was independently developed by the other party without access to your company's confidential information), be sure to review these exclusions to ensure that they don't conflict with the obligations of confidentiality in the agreement or otherwise lessen the vendor's obligation to keep your confidential information confidential.
- Each party should have an obligation to use a high degree of care to keep the other party's information confidential. This standard of care should be at least equivalent to the standard of care that a party uses to protect its own confidential information.
- The duration of the confidentiality obligations in the software license agreement should extend throughout the term of the software license agreement and for at least a number of years thereafter. Trade secret information should be protected as a trade secret for so long as the information is a trade secret or as otherwise prescribed by applicable law.

In certain software license agreements, it is important to describe what the requirements are in the event one of the parties is compelled, by court or government authority, to disclose the confidential information of the other party. In such a case, consider the following terms:

- The receiving party must promptly notify the disclosing party of this requirement.
- The receiving party will not release the confidential information while the disclosing party is contesting or opposing the disclosure requirement.
- The receiving party must cooperate with the disclosing party and provide assistance to the disclosing party with respect to the disclosing party's efforts to prevent the disclosure.
- Any compelled disclosure will not affect the receiving party's obligations with respect to confidential information that is disclosed (i.e., as a result of the disclosure, the information disclosed will not become one of the exclusions to the requirements of confidentiality in the software license agreement).

Additional requirements with respect to confidentiality include the obligation of the receiving party to return or destroy the disclosing party's confidential information upon expiration or termination of the agreement or as otherwise requested by the disclosing party.

- These requirements may include the obligation of the receiving party to provide a notarized written statement to the company certifying the destruction of the disclosing party's confidential information.

- Confidentiality clauses in software license agreements commonly contain an acknowledgment of both parties that, due to the unique nature of confidential information, there is no adequate remedy at law (e.g., money damages) for breach of confidentiality obligations, and therefore, each party would be permitted to seek and obtain equitable relief (e.g., injunction) without the requirement to prove any loss.

Security is an area that is commonly forgotten in software license agreements. However, in situations where a vendor will have access to critical customer information and will be storing that information in its own data center, it is critical that the software license agreement contain terms and conditions in addition to the confidentiality obligations discussed in this chapter. The vendor's requirements in this area commonly include a requirement that the vendor maintain and enforce physical security procedures that are consistent with industry standards and that provide safeguards to protect against the loss, disclosure, and modification of the customer's confidential information.

Maintenance and Support

Maintenance requirements generally define the vendor's obligations with respect to keeping the software current via updates, upgrades, enhancements, new releases, and the like. Support requirements generally define the vendor's obligations to provide technical support for the software, for example, when the software doesn't work in accordance with the specifications or documentation, telephone and on-site support requirements, and the like. When thinking about what provisions are required in a maintenance and support agreement, it is necessary to go back to the original critical considerations, particularly with respect to how critical the application is that your company is licensing. For the most critical applications, your company will require a level of service, such as for responses and resolutions to issues, much greater than for less critical applications where problems don't impact the business as severely. In any maintenance and support provisions in software license agreement, the parties should consider terms such as the following:

- Maintenance and support fees are typically a percentage of the actual license fees that are paid by the customer to the vendor. These can vary widely, but are commonly in the 12%–18% range. Consider locking pricing on support and maintenance for a fixed period of time (e.g., three years) and then negotiating a cap on maintenance and support fee increases over time (e.g., no more than the percentage change in the applicable CPI or 2%).
- Software is generally heavily discounted off of vendors' list prices. Be sure that your company is paying a percentage of what you actually paid for the

licensed software for maintenance and support and not a percentage of list or any other price.

- The parties should consider the support term—generally at least a year with options to renew for multiple additional terms of one year each.
- In many instances, it is appropriate to include service levels with respect to the support that is to be provided by the vendor. This can include, for example, obligations with respect to a vendor's response to a problem and required problem resolution times. Where such service levels are included, consider providing a remedy for your company in the event the vendor does not perform the support as required.
- Consider what training may be required with respect to the licensed software. This will include the location of the training, any travel that is included in the price that your company is paying for the licensed software and the support and maintenance services, how many of your company's personnel can attend the training, the cost to your company if additional personnel want to participate in the training, and requirements with respect to the materials that must be provided by the vendor.
- Ensure that the vendor is required to provide support throughout the term of the agreement and after the term expires as required. Under no circumstances should the vendor be permitted to withhold support, even if there is a dispute between the parties.
- Consider the amount of time that your company will need to implement new releases. Many vendors will require a short window for implementing new releases, and if the customer doesn't implement within the short timeframe, they will fall out of compliance with the maintenance and support terms. The customer should require as much time as necessary (e.g., six months from the date of delivery) to implement new releases so that it can stay current on maintenance and support obligations.

Announcements and Publicity

Consider whether there will be a restriction on the vendor's ability to announce the relationship or list your company on a list of customers. If your company decides that it wants to prohibit publicity with respect to the transaction, ensure that the software license agreement contains a requirement that the vendor will not make any public announcement regarding the relationship and will not use your company's name without your company's prior written consent.

A further complication is that some vendors are now qualifying their pricing with the customer's obligation to participate in a very wide range of marketing and publicity activities. Those activities are frequently intrusive and time-consuming. If the customer declines to participate in those activities, the vendor can require a renegotiation of the contract pricing. These provisions should be strictly avoided.

Term and Termination

The term of a software license is driven by the license terms. A perpetual license is perpetual, while a term license would be for a term agreed upon by the parties. Termination rights generally arise under the following circumstances:

- A party breaches its obligations under the agreement and fails to cure the breach within a specified period of time after the breaching party is notified of the breach.
- A termination for convenience clause giving a party (usually the customer) the right to terminate the agreement at any time and for any reason.
- Partial termination is appropriate in some types of software license agreements. Partial termination would permit the customer an opportunity to terminate only one application or deliverable without terminating the entire agreement.
- Transition services are common and should be included in software license agreements. Upon termination or expiration of the agreement, the vendor would be required to assist the customer in its transition process.

Additional Contract Terms

In any software license transaction, the customer should ensure that the software license agreement that you have negotiated and signed is the only agreement that governs the terms of your company's use of the software and receipt of support, maintenance, and any other professional services associated with the software. In this regard, your company will want to state specifically that no shrink-wrap, click-wrap, or other terms and conditions that are provided with product or software will be binding upon your company, even if those products and services require an affirmative acceptance of those additional terms and even if someone from your company "accepts" those terms. In addition, the following terms are commonly addressed in software license agreements:

- *Force majeure* provisions should be drafted to excuse parties of their contractual obligations only for reasons beyond their control. Software vendors will attempt to broaden *force majeure* clauses by including events that should be within the vendor's control like labor difficulties and issues with telecommunications providers.
- Prohibitions on the customer's right to assign, common in form vendor license agreements, are commonly unfair to the customer since they limit the customer's ability to assign the license to an acquiring or merging entity. From

a revenue stream perspective, this is good for the vendor since any merger or acquisition that affects the customer will result in additional license fees to the vendor. However, this is largely considered an unfair position.

Summary

Software license agreements can range from the very simply to the incredibly complex. A simple license agreement can be for a noncritical application that costs the company very little in terms of dollars and includes very little risk (i.e., if the application doesn't work, there are alternatives that can be quickly implemented without significantly impacting the company's business). More complex license agreements are encountered when the company is licensing a mission-critical software application (e.g., an ERP or CRM system), spending hundreds of thousands or even millions of dollars, and include complicated, long-term, and complex implementation. Whether a software license agreement is simple or complex matters very little when it comes to vendor form software license agreements. Generally speaking, vendor form software license agreements are one-sided and provide very little in terms of protection to the customer. It is critical, no matter the complexity of the transaction, that the customer read the agreement carefully and propose changes that reduce the risk to the customer and align the transaction more closely to the customer's business objectives and requirements.

Chapter 3

Nondisclosure Agreements

CHECKLIST

Form and Type of Agreement

- ☐ Company's form or vendor's form
- ☐ Unilateral (one-way) NDA
- ☐ Mutual (two-way) NDA

Definitions

- ☐ Precise purpose for NDA
- ☐ Definition of "confidential information"

General Requirements

- ☐ Marking requirements
- ☐ Obligation to return and/or destroy
- ☐ Obligations of confidentiality
  - Reasonable care
  - Consistent with internal practices
- ☐ Internal disclosure of information
- ☐ Employees
- ☐ Agents
- ☐ Subcontractors

- ☐ Others
- ☐ Exceptions to confidentiality
  - No fault or wrongdoing of receiving party
  - Received from third party
  - Independently developed
  - Other exceptions
- ☐ Procedure for disclosure for subpoena/court order
- ☐ Opportunity to obtain injunctive relief
- ☐ Notification of potential or actual breach
- ☐ No obligation to disclose
- ☐ No ownership transfer
- ☐ No removal of proprietary notices
- ☐ Protection of intellectual property (IP)
- ☐ Term
  - Intellectual property (IP)
  - Personally identifiable information (PII)
  - Other protection beyond term
- ☐ Information handling requirements
- ☐ Encryption/other protection for highly sensitive information
- ☐ Residual knowledge

Techniques

- ☐ Avoid use of NDA as final/ongoing agreement
- ☐ Avoid commencement of services before definitive agreement
- ☐ Receipt of competitor's information

Overview

Nondisclosure agreements (sometimes called confidentiality agreements) (NDAs) are used in several situations and transaction types to protect information exchanged by the parties to a transaction. Most notably, NDAs are used at the inception of a proposed business relationship to ensure that confidential information disclosed by the parties prior to executing a definitive agreement is protected from unauthorized disclosure. If the parties decide to enter into a definitive agreement (for example, a professional services or software licensing agreement) following their initial discussions, the NDA would commonly be replaced by the confidentiality provisions of the final definitive agreement. In the foregoing example, an NDA is used as an interim agreement to ensure initial discussions and information exchanged by the parties during such discussions are protected by written confidentiality obligations.

NDAs are not usually intended to be used on an ongoing basis to cover a broader relationship between the parties. NDAs may be used on an ongoing basis for employees, contractors, and others, who may not require anything more substantial in the way of contractual documentation to govern their relationship with the business, but this is generally not the case. This chapter does not address the unique issues inherent in employment relationships.

While NDAs can come in all shapes and sizes, they are typically either *unilateral* or *mutual*. Unilateral NDAs (sometimes referred to as one-way NDAs) protect only the information of one of the parties. This type of NDA is used when information will flow in only one direction or when only one of the parties is concerned about protecting its information. Mutual NDAs (sometimes referred to as two-way NDAs) protect the information of both of the parties. Mutual NDAs are used when information will flow in both directions and when both parties are concerned with protecting their information.

In addition to protecting the parties' confidential information, NDAs are a key means of protecting and maintaining the enforceability of trade secret rights. Disclosure of confidential information without an NDA or similar confidentiality obligations can result in irrevocable loss of trade secret protection.

Of the many types of contract documents one may encounter in technology contracting, by and large, NDAs from vendors and customers are usually very similar, using very similar language. This does not mean that oddities cannot occur, but in most instances the language used in one NDA will be very similar to the language found in another NDA. If you encounter truly broad differences, the party drafting the NDA is either trying to address some unusual aspect of their business or is using a form that is not in step with the industry.

Key Considerations

Included below is a summary of the various topics and issues that should be addressed in any NDA. There are several initial considerations in any potential use of an NDA:

- As an initial issue, your company must decide whether to use its form NDA or that of the other party. In many instances, this is not a significant issue. Most NDAs generally cover the same issues and topics, and the differences between your company's NDA and the other party's NDA are frequently trivial or easily resolved through simple revisions. In all cases in which your company will be disclosing highly confidential information or intellectual property (IP), or if unique regulatory issues are involved, the preference is to use your company's form agreement. In such cases, your company's NDA will likely (and should) contain specific terms to address issues that are critical to your company and are likely not included in the other party's form agreement. If you must use the other party's NDA and you will be disclosing highly confidential information or IP, or there are unique regulatory issues involved, be sure to include the specific terms that are critical to your company in order to protect your company's information adequately and appropriately.

- After determining which company's form NDA will be used, your company must determine whether a unilateral or mutual NDA is appropriate. A business relationship in which both parties will be disclosing confidential information and both parties want to protect information disclosed will require a mutual NDA. However, a unilateral NDA is appropriate if information will flow in only one direction or if only one party is concerned with protecting its information. Never use a unilateral NDA protecting the other party if there is any chance that your company will disclose company confidential information to the other party or otherwise as part of the business relationship.
- Unilateral NDAs are also useful to set the tone of the discussions. That is, they are commonly used to articulate to the other side that your company does not want their confidential information and that you are unwilling to assume any obligation to protect their information if they nonetheless choose to disclose it.
- While NDAs are generally straightforward and commonly present few unique issues, they should be given the same level of review as any other legal document.

Essential Terms

Almost all NDAs have certain common provisions.

- Most NDAs are written in terms of a particular "purpose" for which the confidential information will be disclosed. Ensure the scope of that purpose is not overbroad, but includes all desired purposes for which the information is disclosed.
- Review the definition of "confidential information" to ensure that it captures the information your company wants to protect. If unique types of information will be disclosed (e.g., sensitive personal information, source code, marketing plans), the specific categories of information should be expressly identified in the definition of "confidential information." Failure to properly define the confidential information that the NDA is meant to protect could result in the unprotected exposure of your company's confidential information.
- Avoid requirements that your company must mark information as "confidential" or reduce confidential conversations to writing marked as "confidential" within a defined period of time. While such a requirement may be workable

in very limited engagements (e.g., the review of a limited set of documents in a conference room), in most circumstances this requirement is unrealistic and unlikely to be followed. What's more, if the NDA contains such a requirement and your company fails to follow it, that may result in loss of confidentiality protection for confidential information shared with the other party and expose your company to significant risk. In circumstances when your company is asked to execute an NDA, whether mutual or unilateral, in which the other side will be the primary party disclosing information, it may be more favorable to impose a marking requirement on the other side.

- Most NDAs impose an obligation to return or destroy confidential information disclosed by the other party at the end of the term or upon termination of the NDA. While this is not an objectionable requirement, the language should be revised to ensure that the receiving party may retain such confidential information as is necessary to satisfy any document retention obligations imposed on them by law and to ensure the receiving party is not compelled to remove or delete information when it is commercially impracticable to do so (e.g., removing e-mail containing confidential information from old backup tapes). It is important to follow up on the "return or destroy" requirement to ensure the receiving party has actually removed the disclosing party's confidential information from its facilities and systems.
- The NDA should clearly articulate the confidentiality obligations of each party. It should state precisely that the receiving party must at least use reasonable care to protect the confidentiality of the information disclosed, consistent with the manner in which the receiving party protects its own most confidential information.
- Many NDAs are written such that the parties may only disclose information to their "employees." Such a requirement is often unduly limiting. It is better to revise these agreements to permit disclosure of the confidential information to "employees, agents, and subcontractors having a need to know such information," provided that they are bound by written confidentiality agreements at least as protective of the disclosing party as those set forth in the NDA.
- Watch for language in NDAs that could require your company to obtain new NDAs with your company's personnel or contractors. Avoid language that requires such personnel or contractors "to agree in writing to be bound by this agreement" or that they have "agreed to written confidentiality agreements identical to this one."
- Most NDAs contain common exceptions to the prescribed requirements to protect the confidentiality of the information disclosed and identify that information to which the obligations of confidentiality do not apply. For example, such exclusions frequently include information that is made public through no fault or wrongdoing of a party, information that the receiving

party receives from a third party that is not bound by confidentiality obligations with respect to the information disclosed, or information independently developed by the receiving party without using the confidential information of the disclosing party. Beware of NDAs that expand these exceptions, which are already very broad. When your company's most sensitive confidential information is disclosed, consider narrowing these exceptions and adding language that states that it is assumed that the disclosing party's confidential information does not fall within one of these exceptions and that the receiving party bears the burden of proving that the information is within one of the exceptions.

- Most NDAs articulate a procedure to be followed when the receiving party receives a subpoena or court order to disclose the confidential information of the other party. This procedure typically includes providing notice to the disclosing party and cooperating with the disclosing party to avoid or limit disclosure of the confidential information. Revise these provisions to relieve your company from providing notice to the disclosing party if the applicable legal order or process prohibits your company from doing so. Also, add a provision that clearly states that the disclosure as part of a subpoena or court order does not render the information disclosed nonconfidential under one of the exceptions to confidentiality in the NDA.
- All NDAs should include language permitting the disclosing party to "obtain" injunctive relief if the receiving party breaches its confidentiality obligations. NDAs are frequently drafted to permit the other party to "seek" injunctive relief. Note that a party can always "seek" injunctive relief from a court for a breach of confidentiality obligations. If your NDA contains this limitation, expand it by permitting the parties to not only "seek" injunctive relief, but also to "obtain" such relief.
- The NDA should include a clear obligation on the receiving party to promptly notify your company of any potential or actual breach of confidentiality of which it becomes aware.
- All NDAs should clearly articulate that the disclosing party is under no obligation to actually disclose any confidential information. The NDA should clearly state that the decision to make a disclosure is in the disclosing party's sole discretion.

Additional Considerations

Depending on the information to be disclosed and other factors, the NDA may include additional protections. For example, if highly regulated, personally identifiable information may be shared, the NDA may be revised to include language relating to compliance with applicable law and imposing information security obligations on the receiving party.

- The NDA should make clear that neither party is transferring ownership of, or granting any licenses, express or implied, in their confidential information.
- The NDA should include language prohibiting the receiving party from removing any proprietary notices (e.g., trademark and copyright notices) from the confidential information of the disclosing party.
- The NDA should ensure that IP disclosed by a party is adequately protected (e.g., by granting a limited license to use the IP solely in connection with exploring the proposed business relationship). Remember that not all IP is necessarily "confidential information" (e.g., copyrighted material and patented methods may not be confidential). An NDA drafted to protect only "confidential information" may not adequately protect your IP. Where appropriate, make sure that the NDA also protects copyrighted material, patent methods, and other information that may not necessarily be considered "confidential information." For example, consider the following language:

  Nothing contained in this Agreement will be construed as granting the receiving party any right, title, or interest, express or implied, in or to the disclosing party's patent, copyright, trade secret, and other intellectual property rights. Each party expressly reserves all rights in and to its intellectual property.

  If software will be disclosed, appropriate protections regarding reverse engineering, use to develop competing products, etc. should also be included.
- While many NDAs have defined terms (e.g., five years), it is better not to have a term, unless your company is only receiving information and not disclosing confidential information. If it is necessary to include a term, ensure that trade secrets, personally identifiable information, and other highly sensitive information remain protected on an ongoing basis. In particular, some courts have found that trade secrets disclosed under an NDA that has a defined term will lose their trade secret status forever at the end of that term. To address these issues, include language like the following:

  Notwithstanding anything to the contrary contained in this NDA, all personally identifiable information shall be maintained in confidence in perpetuity and all trade secrets disclosed by a party shall be held in confidence for as long as the information is protected by applicable trade secret law.
- Except in limited circumstances, an NDA should not be used as a final, ongoing agreement. Rather, most NDAs should be used as transitional documents in anticipation of a definitive agreement between the parties that contains a robust confidentiality clause.
- Avoid situations in which a vendor or supplier executes an NDA in anticipation of entering into, for example, a professional services agreement, but is then allowed to commence rendering services before the professional services agreement is executed. This is a disfavored practice because the vendor or supplier will be rendering services in the absence of a definitive agreement and the protections contained in such an agreement (e.g., warranties, indemnities, IP ownership provisions). What's more, allowing the vendor or supplier

to commence performance will likely seriously undermine your company's negotiating leverage for the professional services agreement. One means of reducing the possibility that a vendor or supplier will commence performance before execution of the final agreement is to provide a disincentive for commencing work. This can be done through language like the following:

If Vendor elects to commence performance of work for Company prior to execution of a final Professional Services Agreement, Vendor agrees all such work shall be deemed gratuitous and Vendor shall not be due any compensation for the work.

- Care should be taken in any situation where your company will receive confidential information of a party that is or may become a direct competitor or that offers products or services that are similar to your company's products. Receiving such information may "taint" your company's development process.
- If highly sensitive information is at risk and that information will be stored by the receiving party on electronic media (e.g., on computer hard disks, USB fobs, CD ROMs), consider including a specific obligation on the receiving party to irrevocably remove the information from its electronic media using measures consistent with the best industry practices for the permanent deletion of electronic information. Depending on the sensitivity of the data, it may be appropriate to include specific destruction requirements (i.e., at least as protective as the DoD 5220-22-M Standard, NIST Special Publication 800-88, Guidelines for Media Sanitization, or NAID standards). This can be done by way of an "information handling" exhibit attached to the NDA.
- If highly sensitive information is at risk, consider including specific information security requirements to ensure the information is properly secured. For example, sensitive data may be required to be encrypted using specified means, not transmitted over unsecured wireless networks, not placed on removable media, stored on servers that are logically and physically secure.
- NDAs may include language regarding residual knowledge (i.e., knowledge that is retained in the unaided memories of the receiving party's personnel). Such knowledge is commonly not subject to the confidentiality protections in the NDA. Among other things, assessing whether to include a residual knowledge provision in the NDA turns on the type of information at risk, the party placing the most information at risk, and whether the information also constitutes the IP of the disclosing party. Generally, mutual residual knowledge clauses may be appropriate, provided they make clear that residual knowledge may not include, reference, or be derived from any of the disclosing party's confidential information and cannot infringe the disclosing party's IP rights. The language should also make clear the residual knowledge rights do not include any license, express or implied, to the disclosing party's patents, copyrights, trade secrets, or other IP rights. Beware of receiving parties who request deletion of the reference to trade secrets. Deleting that reference would place the disclosing party's trade secrets at substantial risk.

Summary

NDAs are flexible documents that can be used in a broad range of engagements. However, they should not be used for purposes beyond protecting each party's confidential information. There is a tendency to try to turn NDAs into broader agreements that cover services and other obligations. This practice should be avoided in favor of using the NDA only to address confidentiality during the parties' initial discussions and using a properly drafted definitive agreement to govern the relationship between the parties.

Chapter 4

Professional Services Agreements

CHECKLIST

Preliminary Considerations

- ☐ Due diligence process
- ☐ Form of professional services agreement

General Requirements

- ☐ Term—open-ended or limited
- ☐ Termination provisions
  - – Material breach of agreement
  - – Insolvency/cessation of business
  - – For cause
- ☐ Acceptance testing
  - – Procedures
  - – Criteria
  - – Remedies at no cost
  - – Timeframe for failures to correct
  - – Payment of fees linked to milestones and testing completion
- ☐ Personnel requirements
  - – Minimum skill requirements
  - – Interview process
  - – Naming of key personnel

  - Replacement procedures; no charge for overlap and ramp-up
  - Efforts to ensure consistency of personnel
  - Turnover penalties
  - Supplier has sole responsibility for personnel
- ☐ Use and identification of subcontractors
- ☐ Warranties
  - Material compliance of services
  - Compliance with applicable laws
  - Workmanlike manner
  - Timeliness
  - Disabling devices
  - No IP owned by third parties
  - Compliance with data security and privacy laws
- ☐ Indemnification
  - Intellectual property infringement
  - Violation of applicable laws
  - Personal injury and property damage
  - Breach of confidentiality
  - Procedures to permit supplier control of claim
  - No settlement without customer consent
  - Exclusion from limitation of liability
- ☐ Limitation of liability
  - Mutual
  - Exclude indemnity obligations
  - Exclude gross negligence and willful misconduct
  - Exclude confidentiality and data security breaches
  - Tie cap to all fees paid under agreement
- ☐ Intellectual property ownership
  - License to use preexisting materials
  - Who owns IP developed under the agreement
  - If supplier owns, restrictions on use (competitors)
  - Identification of third-party IP and fees associated with third-party IP
- ☐ Change order process
- ☐ Confidentiality and information security
  - Simple provision (basic information)
  - Detailed provision (sensitive information)
  - Information security requirements depending on nature of data
- ☐ Force majeure
  - Ensure proper scope
  - Avoid overbroad provisions to include staffing problems, unavailability of materials, and failure of third parties
  - Right to terminate
  - No payment for services not rendered

- ☐ Nonsolicitation of supplier's employees
- ☐ Insurance tailored to customer's requirements
- ☐ Fees and costs
  - All fees expressed in contract, statement of work (SOW), or change order
  - Payment schedule for all fees
  - Fixed fee vs. time and materials
  - Overall cap for time and materials projects
  - Ensure estimates are accurate
  - Specify percent over estimate to be paid by supplier
  - Specify percent over estimate to be shared by both parties
  - Rate card for future services
  - Allocation of taxes (customer pays only for tax on services received)
  - Payment of fees tied to performance
  - Holdback if payment based on passage of time
  - Travel and expenses tied to customer's policies
  - Financial audit rights

Relationship to Other Agreements

- ☐ All contract terms in a single agreement
- ☐ If multiple agreements, ensure termination rights across agreements
- ☐ Acceptance testing of services linked to acceptance testing of related software and hardware
- ☐ Limitation of liability caps accounts for fees paid across agreements

Overview

Professional services agreements are used for a wide range of service engagements, including software development, implementation work, outsourcing services, website development, and many other activities for which services are rendered for an agreed-upon fee. While these types of agreements are frequently entered into on a stand-alone basis, they are also commonly part of larger engagements. For example, a software license agreement may also include a professional services agreement or component (i.e., professional services terms, or a professional services exhibit, included in a software license agreement) for customization work or implementation services. Similarly, a hardware purchase agreement may include a professional services agreement or component for installation services. Depending on the nature of the engagement, it may be appropriate and more protective in these larger engagements to combine the agreements into a single contract with uniform provisions. In that case, the agreement would be appropriately called, for example, a master license and services agreement or a hardware purchase and installation agreement.

Every professional services contract is composed of at least two parts: the main agreement, containing the legal terms and conditions, and at least one statement of work (SOW) describing the specific services to be rendered by the contactor. In general, the main agreement is written to permit the parties to enter into any number of subsequent SOWs. Each SOW is usually sequentially numbered and specifically references and is incorporated into the main agreement.

While there are many key issues in a professional services contract, two are frequently the most critical: project management and work description. The first relates to ensuring the project has proper oversight to keep the project on target from both a scheduling and a cost perspective. The second relates to the most foundational aspect of professional services. Without a clear description of the specific services to be performed, the other protections in the agreement are largely useless. The service description (i.e., the specific tasks to be performed by both parties, the project schedule, milestones, deliverables, and pricing) all must be set forth in a detailed SOW. Chapter 5 provides a separate discussion of the critical issues involved in drafting SOWs.

As a frame of reference, in the last few years, the number one cause of disputes and costs overruns involving technology contracts have arisen out of poorly drafted SOWs. Pay very close attention to the recommendations in this chapter and in Chapter 5.

Key Considerations

- As an initial consideration, potential contractors should be vetted through a due diligence process to ensure they have the necessary skill set and personnel to perform the required services. This can be accomplished through a formal process, such as a request for proposal (RFP) or request for information (RFI), or informally through discussions with the contractor.
- Due diligence of potential contractors may also extend, depending on the type of services to be rendered, to information about the contractor's information security measures, disaster recovery preparedness, intention to use subcontractors, and staff-turnover rates.
- If information disclosed during the diligence process is key to your decision to select a vendor, the contractor's specific responses may be included as an exhibit to the agreement with a clear statement that you are relying on those responses in entering into the agreement.
- Another key preliminary consideration is whether to use the contractor's form professional services agreement or one of your own. Unlike many other types of vendor engagements, it is not uncommon in professional services engagements for the customer to insist on the use of its own form agreement. The following section covers key provisions that should be considered in each of your company's professional services agreements.

Essential Terms

Term and Termination

Most professional services agreements are written with an open-ended term. That is, the agreement will continue until all SOWs are completed and one of the parties gives written notice of its intent to terminate. In some cases, the parties may also include a minimum base term of one to five years, with optional renewal terms of, for example, one year each. In any case, however, the agreement will not expire while any SOWs are pending.

- Termination rights include the following: termination for material breach of the agreement; the contractor becoming insolvent, ceasing to do business, or becoming the subject of a bankruptcy proceeding; and the contractor assigning the agreement without the customer's authorization.
- Termination for convenience, without cause, may be appropriate for many types of engagements. Ordinarily, these provisions permit the customer to terminate, without penalty or further obligation, on prior written notice (such as thirty calendar days) to the contractor. Any attempt by the contractor to make the provision mutual should be rejected. The parties are not in the same position vis-à-vis each other. A contractor termination right could leave the customer with a half-completed project.

Acceptance Testing

Acceptance testing is one of the most critical protections in a professional services agreement. It is the procedure by which services and deliverables are tested to ensure they materially conform to the specifications agreed upon by the parties in the applicable SOW. In most cases, the acceptance testing procedures are defined in the main agreement, and the specific criteria against which acceptance testing will be measured are set forth in the relevant SOWs. The acceptance testing procedures describe how the parties will test the relevant services and deliverables, how nonconformities will be addressed, and the remedies available to the customer if nonconformities in the services and deliverables are not resolved within a specified period of time (e.g., the customer may terminate the agreement with respect to the nonconforming services and deliverables and receive a refund of fees paid for them). It is important to ensure that work performed by the contractor to remedy its failure to deliver conforming services and deliverables be rendered at no additional charge to the customer. It is also important to include a specific limit on the time a contractor can take in correcting a nonconformity (many contractor form agreements afford the contractor unlimited time to correct a nonconformity).

- Most contractor form agreements lack any form of acceptance testing. Customers should insist on at least some form of acceptance testing in all engagements. The only potential exception would be instances in which the professional services engagement is essentially a contingent workforce or staff augmentation engagement. That is, the contractor is furnishing personnel who will perform services solely at the direction of the customer (i.e., the contractor personnel are acting as supplements to the customer's own workforce). In such cases, the customer essentially controls whether or not the services are successful. This would not, of course, relieve the contractor of liability for work negligently performed.
- Fees should be linked to acceptance testing. In many cases, 20% or more of the overall fees for a particular service or deliverable are withheld until acceptance is achieved. This ensures the contractor will have an appropriate incentive to perform.
- For the protection of both parties, acceptance criteria should be objective and clearly identified in the relevant SOWs.

Personnel

A contractor's performance and the success of the project will turn on the quality of the personnel assigned to perform the services. The contract should require all personnel to have experience in the specific type of services to be rendered. In some cases, it may be appropriate to list minimum skill requirements (e.g., number of similar prior projects, number of years of experience) and to designate certain key vendor personnel, who are critical to the success of the project and cannot be changed without customer consent. In certain instances, it may also be appropriate to interview some or all of the personnel to be assigned to the project and specifically name the project team in the SOW.

- The customer should have the ability to require replacement of contractor personnel who are unqualified, present a security risk, and for any other lawful reason.
- If the contractor must replace personnel during the pendency of a SOW, the customer should not be charged for "ramp-up" time to acclimate the new worker to the project.
- The contractor should be required to use all reasonable efforts to ensure the consistency of its project staff during the term of the SOW.
- If the contractor has a historically high turnover rate (e.g., offshore vendors sometimes have staff turnover rates of 20%–40%), remedies may be included if a specified level of turnover occurs during the pendency of a SOW. High staff turnover has been repeatedly shown to increase cost, create project delays, and create information security risks.
- Ensure the contract makes the contractor solely responsible for the compensation of its workers, their benefits, and immigration status.

Subcontracting

Customers enter into professional services contracts based on the reputation of the contractor they are hiring. If the contractor can immediately hand off performance to one or more subcontractors, the customer may not receive the level of expertise and quality expected. In addition, the use of subcontractors may increase risk (e.g., sending customer data and intellectual property offshore, having services performed by an entity with whom the customer has no contract). If the contractor intends to use subcontractors, the contractor must specifically identify those subcontractors in the SOW before it is executed. After the SOW is signed, the customer should have approval rights over the use of any other subcontractors.

Warranties

Professional services contracts should include baseline warranties ensuring the services materially comply with the requirements of the agreement and associated specifications and documentation, services will be rendered in compliance with applicable law, and services will be rendered in a workmanlike and timely manner. Depending on the nature of the services to be rendered, warranties may be added relating to preventing viruses and other disabling or harmful mechanisms in software, precluding the use of intellectual property owned by third parties without properly obtaining a license from the owner, and compliance with data security and privacy laws.

Indemnification

At a minimum, an indemnity should be included protecting the customer from claims that the contractor's services and deliverables infringe the intellectual property rights of a third party. Common exclusions from this obligation include claims based on specifications provided by the customer, or the customer's combination of the services and deliverables with third-party software or hardware not expressly approved by the contractor. The indemnity should also include a remedy that in the event of an infringement claim, the contractor will be under a duty to provide one of the following protections within a defined period of time, usually two to three months: (i) obtain a license for the customer to continue using the services and deliverables; (ii) modify the services and deliverables, without loss of material functionality, so that they are no longer infringing; or (iii) terminate the agreement with regard to the infringing services and deliverables and issue a refund to the customer of the fees paid for them, usually prorated over three to five years.

- Other common indemnities included in professional services agreements include claims arising from the contractor's violation of applicable laws or regulations, personal injury and property damage (generally included when the contractor will be rendering onsite services at the customer's facilities), and, if sensitive personal information will be at risk, breaches of confidentiality.

- The indemnity provision normally permits the contractor to have sole control over the defense and settlement of any indemnified claims. While this is common, language should be added making clear that the contractor cannot enter into any stipulated judgment or settlement that would bind the customer without the customer's prior written authorization.
- All indemnity obligations must be viewed in relation to the limitation of liability (see the next section). If they are not excluded from all limitations and exclusions of liability, their protection will be greatly limited. It is highly uncommon to place any limitations on a contractor's indemnification obligations for intellectual property infringement.

Limitation of Liability

Almost all professional services agreements will include a limitation of liability, which disclaims certain types of liability and places a cap on all other damages. The presence of such a provision is not objectionable. Certain protections, however, should be included:

- The limitation of liability should be mutual, protecting both parties.
- The contractor's indemnification obligations should be excluded from all limitations and exclusions of liability.
- In appropriate cases, the contractor's gross negligence and willful misconduct should also be excluded from all limitations and exclusions of liability.
- If highly sensitive data will be at risk, breaches of confidentiality and breaches of the contractor's obligations to protect your company's data should also be excluded from the limitation of liability.
- Most limitations of liability are written in terms of excluding all liability for consequential damages and capping direct damages at some portion of the fees paid to the contractor under the agreement. In general, if a multiplier of fees paid is used, the cap should be tied to all fees paid to the contractor, at minimum, under the SOW giving rise to liability. It is preferred to have the cap tied to all fees paid under the agreement. If the agreement will be in place for a long term, the cap may be limited to the fees paid to the contractor during a particular timeframe. For example, the cap may be limited to the fees paid during the preceding twenty-four months. In addition, the cap should ensure at least a minimum level of liability during the early days of the engagement when little in the way of fees may have been paid. This minimum level is usually set at between 50% and 75% of the projected fees for the initial year.

For example, each party's aggregate liability to the other party for all damages, losses, and causes of action whether in contract, tort (including negligence), or otherwise shall not exceed the greater of $100,000 or two times the total fees paid hereunder by licensee during the twelve months preceding the initial event giving rise to such liability.

Intellectual Property Ownership

Another key provision in any professional services engagement is intellectual property ownership. If the contractor will not be rendering services in which intellectual property will be developed, all that may be needed is clear language that any intellectual property furnished by the customer for use by the contractor is subject to a limited license permitting the contractor to use the intellectual property solely for the customer's benefit in connection with the services. On completion of the services or termination of the agreement, the license should automatically terminate. Avoid any language that would require the customer to assign any intellectual property rights to the contractor.

- If the contractor will furnish its preexisting intellectual property to the customer, the customer must have a license to use that intellectual property for, at a minimum, the duration of the professional services agreement. The scope of the license should include all uses the customer intends to make of the intellectual property.
- If the contractor will develop intellectual property during the engagement, then the foundational question is "Who will own that intellectual property?" Depending on the circumstances, the customer may be comfortable having a perpetual license to continue to use the new intellectual property in connection with its business. In other cases, the customer will want to exclusively own the new intellectual property. In those instances, the agreement must clearly state the contractor is assigning all intellectual property rights (not just copyright) to the customer.
- If the customer is willing to only have a license to the new intellectual property, this means the contractor may use that intellectual property for its other customers, including, potentially, competitors of the customer. Some customers may want to place restrictions on such uses (e.g., preventing the contractor from licensing the new intellectual property to specific competitors of the customer or imposing a "cooling-off" period of, for example, eighteen months before the contractor may license to a defined set of the customer's competitors).
- The agreement should also include language requiring the contractor to identify any intellectual property owned by a third party and delivered to the customer in connection with the agreement. The language should make clear

who is responsible for paying any license fees associated with that third-party intellectual property and ensuring that the contractor discloses any third-party contractual terms that may be binding on the customer if it uses that intellectual property.

Change Order

The agreement should include a change order process that requires any changes to the SOW to be expressly set forth in a writing signed by both parties. In particular, all change orders should identify any fee increases or alterations to the project schedule.

Confidentiality and Information Security

Confidentiality protections should be included that are appropriate to the sensitivity of the information to be disclosed. If only basic business information may be shared, a relatively simple confidentiality provision may be used. If highly sensitive trade secrets or personal data may be disclosed, far greater specificity may be required regarding the parties' confidentiality obligations. In some cases, additional provisions may be added regarding information security protections the contractor must implement to ensure the customer's data is adequately secured. Depending on the nature of the data to be shared, applicable laws and regulations may mandate specific information security and privacy protections be included in the contract.

Force Majeure

Typically, force majeure clauses cover natural disasters or other "acts of God," war, and similar occurrences. They become problematic when they are broadened to include staffing problems, unavailability of materials, or the failure of third parties (e.g., suppliers and subcontractors) to perform their obligations to the contracting party. These broader provisions would excuse the contractor for almost any problem. Force majeure clauses should be limited only to major occurrences, such as those mentioned in the first sentence of this section.

- If a contractor's performance becomes the subject of a force majeure event and it cannot perform for a specified period of time (e.g., fourteen days), the customer should have the right to terminate the agreement without further obligation.
- Another important protection is to ensure the customer never pays for services that are not being rendered. This can be accomplished by including language like the following: "In the event Contractor's performance of the Services is the subject of a force majeure event, the fees to be paid by Customer will be equitably adjusted to reflect the period in which Contractor's performance was effected."

Nonsolicitation

Many vendor form agreements include a clause preventing the customer from intentionally soliciting the vendor's personnel to leave the vendor and join the customer. While not generally objectionable, care must be taken to ensure these provisions are limited to vendor employees with whom the customer actually has contact in connection with the services and do not prevent the customer from hiring employees who independently approach the customer or who respond to a general solicitation the customer has made in a newspaper, magazine, or on the Internet. Beware of loosely written provisions that prevent the customer from soliciting not only the contractor's employees, but also their contractors (which might include even large companies with which the contractor does business).

Insurance

Some customers have minimum insurance requirements for their contractors. If so, those requirements should be included. At a minimum, the contractor should be required to carry workers' compensation insurance consistent with applicable law and commercial general liability insurance. Professional errors and omissions coverage may also be appropriate in certain engagements.

Fees and Costs

- The agreement should make clear that the contractor will only be compensated for services that are expressly authorized in a SOW, change order, or other writing from the customer. Without that express authorization, any work rendered by the contractor will be deemed gratuitous and no compensation will be made. This is to ensure the customer is obligated to pay for only what it actually has authorized.
- All fees should be clearly defined in the relevant SOWs, including when those fees become due (e.g., on completion of a milestone, on acceptance of a deliverable).
- As a general rule, fixed-fee engagements are preferred to time and material engagements. In a fixed-fee engagement, the contractor is obligated to complete the services for the agreed-upon fee, regardless of whether the actual fees the vendor incurs in completing the work is in excess of that amount. Time and materials engagements provide for the contractor to be compensated on an hourly or daily basis with no guarantee the work will be completed within any fixed amount.
- Time and materials engagements are frequently used when the exact services cannot be well defined and in instances where the customer will be specifying on an ongoing basis, the exact level of resources it will require from the contractor.

- Time and materials engagements place the majority of risk for achieving a project within budget on the customer.
- If a particular project cannot be well defined from the outset and the contractor is not comfortable proceeding on a fixed-fee basis, the parties may consider entering into an initial "scoping" SOW in which the contractor will be given a defined period of time to better scope the services necessary to complete the project. At the end of the scoping phase, the parties will either agree on a fixed fee, agree on further time and materials work, or terminate the agreement.
- All time and materials engagements require the customer to aggressively manage the project and require frequent reports from the contractor detailing hours spent, fees incurred, and expenses.
- If a time and materials engagement cannot be avoided, all work should be subject to an overall fee cap that cannot be exceeded without the customer's prior written authorization. In addition, except in limited instances (e.g., where it really is impossible for the contractor to predict the level of resources needed to complete a project), the contractor should ensure that the overall cap for the engagement is not a guess, but a good faith estimate based on the contractor's experience and knowledge about the project. Additional protection can be obtained by adding incentive language to ensure the estimate is made in good faith. This can be accomplished by including a risk-sharing mechanism in the event the overall cap is exceeded by a substantial margin. For example, if the contractor exceeds the cap by less than 10%, but completes the work, most customers would agree this result reflects nothing more than the inherent "play" in an estimate. On the other hand, if the cost to complete the project exceeds, say, 20% of the original estimate, this suggests the original estimate was nothing more than a guess. In such a case, the excess cost should be shared by the parties (e.g., each party will pay 50% of the excess fees). In some cases, it may be appropriate to shift the risk of a guess entirely to the contractor. Consider the following example provision:

  Seven hundred hours of Services at the rate of $200/hour represents contractor's best, good faith estimate of the professional services fees required to fully complete all services and deliverables described in this SOW. In the event contractor's aggregate fees exceed the estimate by less than 10%, customer will be responsible for such fees; provided contractor has first obtained customer's prior written authorization to exceed the original estimate. Contractor will bear all professional services fees in excess of 10% of the foregoing estimated amount (e.g., if contractor's fees to complete the services and deliverables are 15% greater than the original estimate, customer will pay the first 10% and contractor will pay the remaining 5%).
- If the contractor will be involved in rendering services over a long period under several SOWs, the customer should require the contractor's rate card of professional services rates be attached as an exhibit to the agreement and

those rates be fixed for at least two to three years. Thereafter, rates may be allowed to increase on a yearly basis, subject to a cap tied to the consumer price index or a simple percentage (e.g., 4% per year).

- The contract should make clear the customer will only be responsible for those taxes based on its receipt of the services (excluding any taxes based on the contractor's revenue or personnel taxes). In some states, there are no taxes on professional services. Consult your tax professionals for counsel on these issues.
  - If the contractor is located offshore (e.g., in India, Russia), local tax laws may change during the performance of the agreement, resulting in potentially substantial increases in the tax on services. In such cases, the agreement should address who bears the risk of those increases (e.g., the risk is entirely on the vendor of any increases during the pendency of the current statements of work, the parties equally share the increase).
- In general, fees should be tied to performance (e.g., achievement of milestones, completion of deliverables, acceptance). Avoid fees tied only to the passage of time. If that cannot be avoided, consider a holdback of 10%–20% of each monthly invoice, payable on final acceptance.
- Travel and living expenses should be subject to the customer's then-current policies for vendor expenses. If the customer does not have a policy, basic language like the following can be used:

  In the event that customer requests contractor to provide services at a location away from the metropolitan area of contractor's regular place of business, customer will reimburse contractor for reasonable travel and living expenses incurred by contractor that would not have been incurred in any event if such services had been performed at contractor's regular place of business. Receipts or reasonable evidence thereof is required for commercial travel, car rental, parking, and lodging. Contractor shall submit monthly expense reports to customer. When contractor employees visit more than one customer on the same trip, the expenses incurred will be apportioned in relation to time spent with each customer. Contractor shall obtain customer's prior written approval, which shall not be unreasonably withheld, before incurring any expenses exceeding, in the aggregate, one thousand dollars ($1,000.00). All air travel shall be coach class on generally scheduled commercial flights. Contractor shall use commercially reasonable efforts to make airline reservations for travel sufficiently in advance of the travel date so as to obtain the lowest airfare.
- To provide the contractor with an incentive for accurate billing, the agreement should include language permitting the customer to audit the contractor's records to ensure bills are correct. The customer generally bears the cost of the audit, unless the audit reveals, for example, overbilling in excess of 10%. In that case, the cost of the audit would shift to the contractor.

Relationship to Other Agreements

- If the professional services agreement is entered into as part of a larger engagement (e.g., as part of a software license or hardware purchase arrangement), the customer should strongly consider requiring all contract terms to be reflected in a single agreement. If that is not desirable or cannot be negotiated, it may be appropriate to link the agreements together with respect to the following:
- A termination of one agreement will result in termination of the other agreements (i.e., cross-default). This will ensure that if a vendor breaches a software license agreement, the customer won't have to continue to pay under a professional services agreement to implement software it can no longer use.
- Acceptance testing of the services is linked to successful acceptance testing of related software or hardware (i.e., cross-acceptance).
- Limitations of liability in the relevant agreements should reflect the level of fees paid under all agreements. That is, if the vendor defaults under the license agreement, but has performed the professional services properly, under most vendor agreements the customer would only be able to recover damages under the license agreement, which would be limited to the fees paid for the software. If the agreements are linked, the liability cap in the license agreement would, potentially, be increased to reflect not only the license fees, but also the implementation fees.

Summary

Professional services agreements are used to govern a wide range of potential services, from software implementation to custom development to outsourcing engagements. The most important elements of these engagements include proper project management and a clear description of the services, deliverables, project schedule, and pricing. Without the foregoing, even the best drafted professional services agreement will provide little protection.

Chapter 5

Statements of Work

CHECKLIST

Preliminary Considerations

- ☐ Due diligence process
- ☐ Form of professional services agreement

General Requirements

- ☐ Term—open-ended or limited
- ☐ Termination provisions
  - – Material breach of agreement
  - – Insolvency/cessation of business
  - – For cause
  - – Acceptance testing
  - – Procedures
  - – Criteria
  - – Remedies at no cost
  - – Timeframe for failures to correct
  - – Payment of fees linked to milestones and testing completion
- ☐ Personnel requirements
  - – Minimum skill requirements
  - – Interview process
  - – Naming of key personnel
  - – Replacement procedures; no charge for overlap and ramp-up

- Efforts to ensure consistency of personnel
- Turnover penalties
- Supplier has sole responsibility for personnel

☐ Use and identification of subcontractors

☐ Warranties
- Material compliance of services
- Compliance with applicable laws
- Workmanlike manner
- Timeliness
- Disabling devices
- No IP owned by third parties
- Compliance with data security and privacy laws

☐ Indemnification
- Intellectual property infringement
- Violation of applicable laws
- Personal injury and property damage
- Breach of confidentiality
- Procedures to permit supplier control of claim
- No settlement without customer consent
- Exclusion from limitation of liability

☐ Limitation of liability
- Mutual
- Exclude indemnity obligations
- Exclude gross negligence and willful misconduct
- Exclude confidentiality and data security breaches
- Tie cap to all fees paid under agreement

☐ Intellectual property ownership
- License to use preexisting materials
- Who owns IP developed under the agreement
- If supplier owns, restrictions on use (competitors)
- Identification of third-party IP and fees associated with third-party IP

☐ Change order process

☐ Confidentiality and information security
- Simple provision (basic information)
- Detailed provision (sensitive information)
- Information security requirements depending on nature of data
- Force majeure
- Ensure proper scope
- Avoid overbroad provisions to include staffing problems, unavailability of materials, and failure of third parties
- Right to terminate
- No payment for services not rendered

- ☐ Nonsolicitation of supplier's employees
- ☐ Insurance tailored to customer's requirements
- ☐ Fees and costs
  - All fees expressed in contract, SOW, or change order
  - Payment schedule for all fees
  - Fixed fee vs. time and materials
  - Overall cap for time and materials projects
  - Ensure estimates are accurate
  - Specify percent over estimate to be paid by supplier
  - Specify percent over estimate to be shared by both parties
  - Rate card for future services
  - Allocation of taxes (customer pays only for tax on services received)
  - Payment of fees tied to performance
  - Hold back if payment is based on the passage of time
  - Travel and expenses tied to customer's policies
  - Financial audit rights

Relationship to Other Agreements

- ☐ All contract terms in a single agreement
- ☐ If multiple agreements, ensure termination rights across agreements
- ☐ Acceptance testing of services linked to acceptance testing of related software and hardware
- ☐ Limitation of liability caps accounts for fees paid across agreements

Overview

Statements of work, commonly referred to as SOWs, are the heart of any professional services engagement. It is critical that the SOW clearly describe all services and deliverables the customer expects to receive. Without that clarity, the customer may not receive what they expect, cost overruns may result, and delivery schedules may be missed. Remarkably, even though SOWs are key to the success of a professional services engagement, they are frequently written at the last minute, seldom given the time they deserve to perfect, and generally fail to provide the level of detail necessary to clearly define the services and deliverables to be provided. SOWs are also frequently drafted by contractors with little or no input from the customer's business.

SOWs are typically used as exhibits to an overarching agreement of some kind (e.g., a professional services agreement, implementation agreement, software development agreement) to define specific requirements for a set of services or project. The SOW usually includes a detailed description of the services to be provided, deliverables, milestones, payment terms, service levels, acceptance criteria, specifications, and due dates. In most cases, the SOW describes "what" the contractor will be doing, "where" the contractor will be doing it, "who" will be providing the services, "when" the services will be provided, and "how much" the services will cost. The SOW does not address legal terms, which should be addressed solely in the agreement. However, SOWs often include additional commercial terms that supplement the commercial terms in the agreement. Generally, commercial terms that may be included in a SOW should *never* supersede conditions defined in the agreement and should only be permitted in a SOW where the agreement expressly contemplates these additional or different commercial terms.

In the typical scenario, the contractor usually prepares a proposal for the scope of work to be performed, and the agreement should include a description of the type of information that must be included in each SOW (e.g., specific services to be performed, deliverables, project schedule, acceptance criteria, fees, expenses). The contractor's proposal should not replace the SOW; with appropriate revisions, it can, however, be incorporated into the SOW, perhaps as an attachment. Once the customer approves the proposal, the parties typically prepare a draft SOW, which is then reviewed by the project manager and all other relevant stakeholders (e.g., legal, compliance, information security) prior to signing.

The importance of ensuring the SOW clearly describes all services and deliverables to be provided by the contractor cannot be overemphasized. Courts have made clear that the contractor is only responsible for delivering those items expressly identified in the SOW. If a deliverable service or any other SOW element isn't clearly and expressly identified, the contractor will not be obligated to provide it.

If the SOW cannot be negotiated and executed concurrently with the agreement, the agreement should be clearly drafted to ensure the customer has no payment obligation and may terminate the agreement if a SOW is not agreed upon. That is, if completion of the SOW must be deferred until after execution of the agreement, the customer should not be required to commit to a purchase of services until the SOW is fully negotiated. Doing otherwise would seriously undermine the customer's negotiating leverage and potentially obligate it to pay for services without an acceptable SOW in place. The foregoing is particularly important in software license agreements, where the customer may be required to irrevocably commit to pay the license fees, while the critical SOW to implement the software is left for future negotiation and agreement.

The following is a checklist of the various topics and issues that should be considered in drafting SOWs. The checklist should assist with the review of a SOW to ensure that it is consistent with the requirements of the underlying agreement and with the customer's legal and contract management requirements and processes.

Essential Terms

Scope of Work and Business Requirements

This is the portion of the SOW that describes the scope of the services to be provided and details the customer's business requirements.

- Include an overview section at the beginning of each SOW providing a brief, plain-English explanation of what the contractor is expected to do or deliver. This explanation should be written so that someone unrelated to the project who is generally familiar with technology could understand the services and deliverables to be provided by the contractor. Avoid excess use of jargon and acronyms and capitalized terms, unless such terms are clearly defined in the SOW or the agreement.
- Include a project plan with a clear project schedule. All dates must be readily calculable. Avoid referring to dates as "estimates." Avoid calculation of all dates from the "beginning of the project" without clearly defining the date of that beginning. Where appropriate, include credits for failure to adhere to the project schedule (e.g., credits to be issued for each day/week a deliverable is delayed). Credits should scale according to the length of the delay. Consider "earn-back" language to permit the contractor to earn back the credits if it promptly returns to the required schedule.
- Functional and technical specifications are essential. Avoid deferring the specifications to a later date. If part of the project includes the development of detailed technical specifications later in the project process, include a description of the requirements for the development of specifications.
- Avoid using unmodified contractor form implementation plans. Each implementation effort is unique to the customer and should be modified to reflect the specifics of the engagement.
- Remove or limit extensive lists of contingencies on the contractor's performance. Carefully review and limit any contractor "assumptions" in the SOW. The vast majority of those contingencies are very general in nature and would create a substantial "out" for the contractor and provide the means for the contractor to charge additional fees. Often, the assumptions should be reevaluated as requirements. For example, if the contractor assumes that the customer will provide all workstations and development software, this statement should be recast as a requirement for the project and include detailed descriptions of exactly what software or hardware is required and when it is required.
- Avoid references to an associated proposal, request for proposal ("RFP"), or other sales-related documentation. These documents may contain legal terms that could conflict with the agreement. When specific content in these

documents is relevant to a particular project, it should be revised accordingly and then directly incorporated into the SOW.

- Remove language that would allow the SOW to override the agreement, especially when there are conflicting terms. Unless specifically approved by the customer's contract management and legal department, there should be no legal terms in the SOW. In particular, language relating to ownership of intellectual property, warranties, indemnities, and limitations on each party's liability should not be included in the SOW. This language should be fully addressed in the agreement.
- Ensure the language in the SOW conforms to the agreement. This means making sure defined terms used in the agreement are also used in the SOW. Examples of potential problems include the following: referring to the contractor by a name that conflicts with the defined name in the agreement and failure to use defined terms like "Deliverables" and "Services" (most SOWs refer to the "work" to be performed, but the agreement likely focuses entirely on the "Services," a defined term, to be performed).
- Avoid passive voice. Never have statements like the following: "the router will be configured" or "the functionality will be tested." Passive voice means the party doing the acting is not expressed. That is exactly the kind of language no SOW should contain. SOWs should be comprised of clear, affirmative, active voice: "the Vendor will configure the router" and "Vendor will complete the functionality and Customer will test the functionality." More than one dispute has arisen from the use of passive voice. Avoid it.
- Delete all language that would result in deliverables or services being "deemed accepted" simply because of the passage of time. The express procedure and criteria for acceptance should be set forth in the agreement and should not be overridden in the SOW.
- If capitalized terms and/or acronyms are used, they must be clearly defined. If highly technical terms are used, consider including a brief definition.
- Consider breaking complex SOWs into smaller, more discrete SOWs. When necessary, an initial, limited "scoping" SOW may be used to better define the requirements for a later SOW (e.g., it may be necessary for the contractor to perform certain initial assessment services to better understand the customer's systems before it can commit to a SOW for a software implementation). Scoping SOWs may also be useful in changing a time and materials assignment into a fixed-fee engagement. For example, if the contractor has insufficient information about an engagement, it may be reticent to provide services on a fixed-fee basis. However, if the contractor is afforded the opportunity to conduct an assessment of the customer's environment during a brief scoping SOW, it may be more inclined to commit to a fixed-fee engagement.

Technical Environment

Depending on the engagement, it may be important to describe in detail the technical environment in which the services will be rendered or in which the deliverables are expected to operate.

- Define the hardware and software requirements of the project.
- Identify relevant applications.
- Define required interfaces, data formats, and connectivity requirements.
- Specify any other relevant systems and interfaces.

Acceptance Testing

Acceptance testing ensures the services and deliverables conform to the customer's expectations, which are described in detail in the body of the SOW. While the agreement will generally include language defining the procedure for acceptance testing (e.g., when and how acceptance testing will be conducted, procedures for addressing nonconformances, and remedies for failed acceptance testing), the SOW provides the specifics of the testing or the acceptance criteria.

- The period of acceptance testing and the acceptance criteria should be identified. The goal is to objectively define all criteria by which acceptance is to be measured. While acceptance tied to the customer's "reasonable satisfaction" is certainly useful, it would require a lawsuit to determine the meaning of that requirement. It is better to clearly and objectively define acceptance criteria than to use subjective language that may give rise to disputes.
- Include staged acceptance where relevant (e.g., initial installation acceptance, system integration acceptance, end user acceptance, final system acceptance).
- Include final, overarching acceptance where relevant.

Deliverables

Deliverables are the specific items that the contractor is expected to create and deliver as a result of the services. They can be pieces of software, hardware components, or items of documentation.

- What will be delivered and when? Create several of these milestones throughout the life of the project.
- In what form/format will the deliverables be provided?
- Payments should be tied to objective milestones and deliverables, not only date-based terms.

- Be aware of language that shifts all of the risk associated with a delay in the project to the customer. For example, if the contractor requires input or other items from the customer, the customer's failure to deliver those items on time should not create an "out" for the contractor unless the contractor expressly informs the customer of the delay and the likely consequences of the delay. Using this approach allows the customer to control the timeline and forces the contractor to take a proactive approach to managing project timelines, thereby reducing the chance of unwelcome surprises for the customer.
- Documentation should be provided for all deliverables. The description of each deliverable should be specific and detailed enough that in the event of early termination of the SOW, for whatever reason, the work to-date can be leveraged.

Documentation

In certain engagements, it will be important for the contractor to deliver documentation (e.g., user guides, software flow charts, specifications) as part of the deliverables. Documentation should be clearly written and understandable by the audience for which it is intended.

- What will be provided to the contractor?
- What is expected from the contractor?
- In what format will documentation be provided?
- Delete all legal terms included in contractor-provided documentation. Only the agreement should contain legal terms unless specifically agreed otherwise by the customer's lawyers.

Roles and Responsibilities of the Parties

Few professional service engagements require the services of only the contractor. In most instances, the customer will also be assigned tasks to perform. Frequently, the customer tasks will be written as contingencies on the contractor's ability to perform.

- Specify names, titles, and roles for each member of the customer team.
- Specify names, titles, and roles for each member of the contractor team.
  - Identify any key contractor team members (e.g., the contractor's project manager). Include language limiting the contractor's ability to replace key team members.
  - Where relevant, include incentives to ensure consistency of the contractor's team (e.g., a credit if attrition exceeds a defined percentage of the team, which is particularly relevant in offshore engagements where staff turnover is frequently very high).

Project Management Processes

In any project lasting more than a few days, it is important to detail the methods by which the project will be managed to ensure schedules and budgets are achieved and services performed properly.

- What project management tool, if any, will be used?
- Frequency and types of reports. Define content and metrics (service level agreements, budget, milestones, etc.) for the reports.
- Project plan (activities, resources, timeframes).
- Ensure all relevant dates are properly calendared and tracked.
- Progress reports (include tasks completed that week, to be completed the following week, issues and status of deliverables, timesheet, etc.). Agree on the form of the reports.
- Method of updates (meetings, conference calls, video conferences, etc.).

Issue Resolution and Escalation Procedures

The potential for disputes is present in every professional services engagement. The SOW should detail how problems will be escalated to ensure prompt resolution.

- Define the escalation process to be followed in the event of a dispute and parties who need to be involved.
- Include specific timeframes for escalation and resolution.
- Where appropriate, include mechanisms to ensure the contractor will cooperate with joint dispute resolution in multi-vendor environments.
- Tie this process to any dispute resolution process included in the agreement.

Risks

The SOW should include, where appropriate, a listing of relevant risks and requirements for the mitigation of those risks.

- Describe relevant risks (e.g., sensitive data at risk, critical delivery dates, dependencies).
- Define the level of risk (high, medium, or low).
- Define the category of the risk (resources, testing, etc.).
- Define the mitigation plan for each risk (what will be done, what the backup plan is, etc.).
- Are there any unique security risks posed by the contemplated SOW? If so, have the relevant stakeholders within the customer's organization been contacted and is these an agreed-upon plan to mitigate these security risks?
- Is additional due diligence required? If so, what is the scope and which party is doing it?

Pricing and Cost

The SOW should detail exactly what fees and costs are to be paid and when. Language should be included to ensure no hidden costs are present and that budgets cannot be exceeded without prior written authorization.

- Total project cost should be based on negotiated rates. The negotiated rates should be fixed for the duration of each project.
- Time and materials projects are generally disfavored. Fixed-fee arrangements are preferred. If time and materials must be used, ensure the "estimate" provided is not a guess. Include a "not-to-exceed" amount. Estimates should clearly state they are based on the contractor's best, good faith belief of the fees required to complete the services, which is based on the contractor's experience in rendering services of this kind. Consider mechanisms for sharing of risk if the original estimate falls far short of the actual fees (e.g., if final fees are more than 10% greater than the original estimate, and the excess is not attributable to the customer's actions or inactions, the excess fees should be split 50/50 between the parties; in this way, the contractor shares the risk of providing a "guess," rather than an educated estimate).
- Define and, if necessary, negotiate reimbursable expenses (capped whenever possible). All expenses should be subject to the payment terms in the underlying agreement.
- Define the invoicing schedule and payment terms. The agreement should include general terms and conditions associated with invoicing requirements and payment terms. Typically, contractors should bill on a monthly basis for time and material engagements. For fixed-fee projects, milestones and deliverables should be used to establish billing timeframes, which should be included in the SOW. Consider holdbacks (e.g., 10%–20% of each invoice) until final acceptance at the conclusion of the project or major milestone or phase.
- Associate all payments with objective project milestones or deliverables. Except in certain time and materials engagements, avoid linking payments solely to the passage of time.
- Include a cap/limit on fee increases for future professional services (e.g., initial negotiated rates are fixed for two years and then may increase no more than by the CPI, Employment Cost Index (ECI), or a fixed percentage).
- Include a cap or limit on fee increases for ongoing support, if applicable. In addition, include specific renewal rights for support.
- Review, identify, and limit all possible revenue streams under the SOW. That is, identify all possible circumstances under which the contractor could charge additional fees and ensure those circumstances are controlled and that additional fees must be preapproved in writing.

Service Level Agreements

Service level agreements ("SLAs") describe the specific performance levels to be achieved by the contractor. SLAs are not relevant in every engagement. For example, SLAs would typically not be required for services relating to the development of product documentation, but would be critical in ensuring hosting services are available when needed.

- Include expectations, metrics, and measurement period.
- SLAs and the associated remedies must be very clearly defined and capable of objective verification.
- Define the contractor's obligations for reporting of SLA performance.
- Include additional remedies for repeated failures (e.g., root-cause analysis, additional termination rights, heightened credits).

Change Orders

During the course of performance of the services, it may become important to the customer to make changes to the services being rendered. Those changes are made through a formal process known as a "change order." Having a clear and documented change order process will reduce the likelihood of cost overruns and project schedule slippage. They also ensure that every modification to the services is fully documented.

- How will changes to business requirements and work be handled?
- When do they affect deliverables and costs of the project? The agreement will generally include general terms and requirements for change orders.
- Beware of contractors that use multiple change orders to "make-up" for underbidding a project.
- Beware of change orders that modify the original scope of the project in such a way that the initial assessment of the risks or business case may no longer be valid. In these instances, the original stakeholders that reviewed the SOW should also review the proposed change order. In some instances, additional due diligence may be required.

Summary

While having an excellent overarching agreement in a professional services engagement is clearly important, those legal terms are meaningless without a properly drafted SOW. Businesses must develop procedures to ensure these key documents are given the attention they deserve and not treated as an afterthought to the negotiation. By developing a checklist for SOWs based on the content of this chapter, revised to reflect the business's own unique requirements, and requiring personnel responsible for drafting SOWs to use that checklist in every engagement, businesses can greatly decrease the risks presented in professional services engagements and have far greater confidence that they will receive the services and deliverables they expect.

Chapter 6

Cloud Computing Agreements

CHECKLIST

Service Levels

- ☐ Uptime
- ☐ Response time
- ☐ Problem response and resolution
- ☐ Remedies

Data Security

- ☐ Protection against security vulnerabilities
- ☐ Disaster recovery and business continuity requirements
- ☐ Frequency of data backups
- ☐ Use of/return of data
- ☐ Format for return of data
- ☐ Review of security policies
- ☐ Physical site visit
- ☐ SSAE 18
- ☐ Limitations on right to use data

Insurance

- ☐ Cyber liability policy
- ☐ Technology errors and omissions
- ☐ Electronic and computer crime
- ☐ Unauthorized computer access
- ☐ Avoid only general liability policy

Indemnification

- ☐ For breach of confidentiality and security requirements
- ☐ For infringement claims
- ☐ No limitation on types of IP covered
- ☐ Consider limitation to US patents

Limitation of Liability

- ☐ Application to both parties
- ☐ Exclusions (from both consequential exclusion and cap on direct damages)
  - Breaches of confidentiality
  - Claims for which the vendor is insured
  - Indemnification obligations
  - Infringement of IP rights
  - Breach of advertising/publicity restrictions
- ☐ Overall liability cap as a multiple of fees

License/Access Grant and Fees

- ☐ Broad permitted use
- ☐ Avoid limitation to internal business purposes
- ☐ Application to affiliates, subsidiaries, outsourcers, and others
- ☐ Consider pricing

Term

- ☐ Free ability to terminate
- ☐ Consider limited notice period
- ☐ Consider limited termination fee (if justified by vendor's upfront costs)

Warranties

- ☐ Data security
- ☐ Redundancy/disaster recovery/business continuity

- ☐ Performance in accordance with specifications
- ☐ Services provided timely and in compliance with best practices
- ☐ Provision of training as needed
- ☐ Compliance with laws (both the software and personnel)
- ☐ No sharing of client data
- ☐ Software will not infringe
- ☐ Software will not contain viruses
- ☐ No pending/threatened litigation
- ☐ Sufficient authority

Publicity/Use of Trademarks

- ☐ No media announcement unless agreed
- ☐ No use of customer marks without permission

Notification for Security Issues

- ☐ Customer gets sole control over notification
- ☐ Reimbursement for costs and expenses

Assignment

- ☐ Ability to assign freely
- ☐ Assignee assumes responsibilities under the agreement

Pre-Agreement Vendor Due Diligence

- ☐ Questionnaire to vendors to include questions regarding
  - Financial condition
  - Insurance
  - Existing service levels
  - Capacity
  - Physical and digital security
  - Disaster recovery and business continuity processes
  - Redundancy
  - Ability to comply with applicable laws

Key Considerations and Essential Terms

Cloud computing is the use of the Internet or other telecommunications links to provide a user with access to software or other technology resources made available at a remote location. Depending on the type of information technology (IT) capability being offered as a service in the "cloud," cloud computing is known by and commonly encompasses several different types of services such as Software as a Service (SaaS), Infrastructure as a Service (IaaS), and Platform as a Service (PaaS). Regardless of the terminology used, cloud computing involves accessing software and infrastructure remotely and frequently includes storing data, often very sensitive and regulated data, in the cloud. While cloud computing agreements have some similarity to traditional software licensing agreements, they have more in common with hosting or application service provider agreements.

When drafting and negotiating cloud computing agreements, it is essential to understand how the application or platform will be used. A good place to start is by comparing the cloud computing model to the classic licensing model for delivery of software. In a traditional software licensing engagement, the vendor installs the software in the customer's environment. The customer has the ability to have the software configured to meet its particular business needs, and the customer generally retains control over the data that is stored in and processed by the software and the system. In a cloud computing environment, the software and the customer's data are hosted by the vendor, typically in a shared environment (i.e., many customers share the same server to access the software, and, therefore, the customer's data is stored by and processed on the same server as other customer's data) and the software configuration is much more homogeneous across all of the vendor's customers (i.e., cloud applications are frequently not customizable or have a limited about of customization available). Accordingly, the customer's top priorities shift from configuration, implementation, and acceptance in the traditional software licensing model to service levels (availability, responsiveness, and remedies) and data (security, redundancy, and use) in the cloud model. It is this reason that traditional software license agreements are not the best framework for cloud computing agreements. However, like a traditional software licensing agreement, a cloud computing agreement will include common provisions such as insurance, indemnity, limitations of liability and warranties, all of which remain important in cloud transactions. When drafting cloud agreements and working through cloud transactions, it is important to note that vendors often refer to "software" whenever discussing the cloud offering, though a particular offering may include services, infrastructure or software, or a mixture of all those. For simplicity, this chapter refers to "software" in the same context.

Service Levels

One of the most critical aspects in drafting and negotiating a cloud computing agreement is establishing appropriate service levels in relation to the availability and responsiveness of the software. Because the software is hosted by the vendor without the customer's control, service levels serve two primary purposes. First, service levels assure the customer that it can rely on the software in its business and provide appropriate remedies if the vendor fails to meet the agreed service levels.

Second, service levels provide agreed-upon benchmarks that facilitate the vendor's continuous quality improvement process and provide incentives that encourage the vendor's diligence with respect to addressing issues. In cloud computing contracts, customers should generally focus on uptime, response time, and problem resolution.

Uptime Service Level

Because the software your company is using is hosted by the vendor, the vendor needs to provide a stable environment in which the software is available to the customer when required, if not twenty-four hours per day, seven days per week. An uptime service level will address this issue by having the vendor agree that the software will have an uptime (i.e., availability) of a certain percentage, during certain hours, measured over an agreed-upon period. By way of illustration, here is an example of this type of provision:

> Vendor will make the software Available continuously, as measured over the course of each calendar month period, an average of 99.99 percent of the time, excluding unavailability as a result of Exceptions, as defined below (the "Availability Percentage"). "Available" means the software shall be available for access and use by Customer. For purposes of calculating the Availability Percentage, the following are "Exceptions" to the service level requirement and the software shall not be considered unavailable if any such unavailability is due to: (i) Customer's acts or omissions; (ii) Customer's Internet connectivity; or (iii) Vendor's regularly scheduled downtime (which shall occur weekly, Sundays, from 2:00 a.m. to 4:00 a.m. Eastern time).

While the specific service level targets (e.g., percentage availability requirement, measurement period, exceptions to availability) depend on the facts and circumstances in each case, an uptime service level (and some form of the example above) is common in cloud transactions. Customers should not simply accept the default vendor positions on uptime percentages, measurement periods, and exceptions, but should instead negotiate terms that address the customer's business needs and requirements and are appropriate for the type, complexity, and importance of the software being used.

Moreover, customers should carefully consider the permitted outage measurement window (e.g., daily, monthly, and quarterly). Vendors tend to want longer measurement periods because they dilute the effects of a downtime and make remedies less available to the customer. Customers should receive written documentation of a vendor's scheduled downtime and ensure the window creates no issues for the customer's business (e.g., is not scheduled during a time period where the customer expects usage of the software). Customers may also request that the vendor be proactive in detecting downtime by explicitly requiring the vendor to constantly monitor the "heartbeat" of all its servers through automated "pinging." Requiring the vendor to do this should result in the vendor knowing very quickly that a server is down without having to wait for a notice from the customer. Finally, the concept of "unavailability" should also include severe performance degradation and inoperability of any software feature. This is discussed in more detail in the next section.

Beware of three recent vendor trends in service level calculations. First, they change the measurement period of monthly to quarterly—thus providing only three opportunities per year to obtain credits and exercise other rights for failure to achieve required service levels. Second, vendors include language stating that an episode of unavailability must persist for at least ten minutes to be counted toward service level performance. This means the vendor can be down an unlimited number of times for nine minutes and fifty-nine seconds every day and the customer will receive no service level credit or other remedies. Finally, vendors are changing how "availability" is defined so that 100% of the functionality of the platform or application may be unavailable, but if their system responds to a basic ping, as described in the preceding paragraph, they will not be considered unavailable. This would greatly undermine the value of any service level.

Response Time Service Level

Closely related to and often intertwined with an uptime service level is a response time service level. This service level sets forth maximum latencies and response times that a customer should encounter when using the software. Remote software that fails to provide timely responses to its users is effectively unavailable. As with the uptime service level, the specific service level target depends on the facts and circumstances in each case, including the complexity of the transaction at issue and the processing required. Note that response time service levels are typically included with the definition of "availability" set forth in the uptime service level section. Here is an example provision of a response time warranty:

> Vendor guarantees that 98 percent of software transactions will exhibit two seconds or less response time, defined as the interval from the time the user sends a transaction to the time a visual confirmation of transaction completion is received.

Problem Resolution Service Level

In addition to the appropriate service levels, a vendor must be obligated to resolve issues in a timely manner. Vendors often include only a response time measurement, meaning the time period from when the problem is reported to when the vendor begins working to address the issue. These obligations typically fall short of what is necessary since they only include a response without any commitment to actually correct the problem. As a result, the service level should include both an escalation matrix (defining the level of severity of the problem and response times for each level) and specific vendor obligations to address and correct the problem or provide an acceptable workaround.

Remedies for Service Level Failure

Remedies should cover both a failure to hit a service level and a failure to timely resolve a reported support issue. Typically, these remedies start out as credits toward the next period's service. For example, a remedy might provide: for every percentage of downtime below the agreed-upon level in the measurement period, or for every severity level 1 support issue vendor does not resolve within the stipulated time, the customer will receive a 5% discount on the next month's bill, up to a maximum credit of 100%. Another format that is commonly used is the following: If the availability service level is 99.99% in each month, customer will receive a $500 credit in each month where the availability is between 98% and 99.98%, and a $1,000 credit if the service level falls below 98%. The remedies should scale such that if repeated failure occurs, the customer should have the right to terminate the agreement without penalty or having to wait for the current term to expire. Here is a portion of a sample remedy provision for a service level failure (the provision for a support failure could be similarly drafted):

> In the event the software is not Available 99.99 percent of the time, but is Available at least 95 percent of the time, then in addition to any other remedies available under this Agreement or applicable law, Customer shall be entitled to a credit in the amount of $500 each month this service level is not satisfied. In the event the software is not Available at least 95 percent of the time, then in addition to any other remedies available under this Agreement or applicable law, Customer shall be entitled to a credit in the amount of $1,000 each month this service level is not satisfied. Additionally, in the event the software is not Available 99.99 percent for (a) three (3) months consecutively, or (b) any three (3) months during a consecutive six (6) month period, then, in addition to all other remedies available to Customer, Customer shall be entitled to terminate this Agreement upon written notice to Vendor with no further liability, expense, or obligation to Vendor.

Data

The vendor's use of customer data and the security and confidentiality of that customer data are very important in cloud computing agreements. The vendor should provide detail regarding and agree to reasonable provisions addressing its competency, policies, and procedures related to: (i) protection against security vulnerabilities; (ii) disaster recovery and business continuity, (iii) data backups; and (iv) the use of, and return of, customer data.

Data Security

The need for data security in cloud computing transactions cannot be understated. While it might seem that cloud computing vendors would want their agreements to include detail about their data security, they too often do not. Accordingly, customers should demand that vendors provide specific details in the agreement about data security, specifically hardware, software, and security policies. These details need to be reviewed by someone competent in data security—either someone within the customer's organization, a data security attorney, or a third-party consultant. Some vendors will not distribute copies of their security policies but will allow customers to come to the vendor's site and inspect them. Such policy inspection should be done if the customer information at issue is very sensitive or mission-critical. Customers should compare the vendor's policies to their own, and in some circumstances it is appropriate for a customer to demand that the vendor match the customer's policy. Verification of the vendor's capabilities with respect to data security, via a physical visit, SSAE 18 audit (IT internal controls audit) conducted by a third party, is also commonly appropriate. It is becoming far more expected that vendors regularly demonstrate to their customers that their security controls remain intact and robust.

Consider the following sample of a typical data security provision:

1. **In general**. Vendor will maintain and enforce safety and physical security procedures with respect to its access and maintenance of customer information that are (i) at least equal to industry standards for such types of locations where customer information will be located, (ii) in accordance with reasonable customer security requirements, and (iii) which provide reasonably appropriate technical and organizational safeguards against accidental or unlawful destruction, loss, alteration, or unauthorized disclosure or access of customer information and all other data owned by customer and accessible by vendor under this agreement.
2. **Storage of customer information**. All customer information must be stored in a physically and logically secure environment that protects it from unauthorized access, modification, theft, misuse, and destruction. In addition to the general standards set forth above, vendor will maintain an adequate level of physical security controls over its facility. Further, vendor will maintain an adequate level of data security controls. See Exhibit A for detailed information on vendor's security policies protections.
3. **Security audits**. During the term, customer or its third-party designee may, but is not obligated to, perform audits of vendor's environment, including unannounced penetration and security tests, as it relates to the receipt, maintenance, use, or retention of customer information. Any of customer's regulators shall have the same right upon request. Vendor agrees to comply with all reasonable recommendations that result from such inspections, tests, and audits within reasonable timeframes.

Disaster Recovery and Business Continuity

Disaster recovery and business continuity provisions require the vendor to demonstrate and promise that they can continue to make the software available even in the event of a disaster, power outage, or similarly significant event. Too often the customer does not request these provisions or, even if they do, they do not read the actual vendor policies and procedures with respect to disaster recovery and business continuity. This is a mistake because customers generally won't have their own up-to-date backup of the data used with or processed by the software. Without access to such data and software on an ongoing basis, even during a disaster, the customer's business may falter. The customer should, therefore, require contractual assurance regarding disaster recovery and business continuity. By way of illustration, here is a sample provision of what to ask for from the vendor in this regard:

> Vendor shall maintain and implement disaster recovery and avoidance procedures to ensure that the software is not interrupted during any disaster. Vendor shall provide Customer with a copy of its current disaster recovery and business continuity plan and all updates thereto during the Term of this Agreement. All requirements of this Agreement, including those relating to security, personnel due diligence, and training, backup, and testing shall apply to the Vendor's disaster recovery site.

Data Redundancy

Because the customer relies on the vendor as the custodian of its data, cloud computing agreements commonly contain explicit provisions regarding the vendor's obligations to back up customer data and the frequency of that backup (e.g., frequent partial backups and periodic full backups). A good place to start is for the customer to compare the vendor's backup policies to its own backup requirements and ensure that the two policies are consistent in all material and critical respects. Below is a sample provision addressing these obligations:

> Vendor will: (i) execute (A) nightly database backups to a backup server, (B) incremental database transaction log file backups every thirty minutes to a backup server, (C) weekly backups of all hosted Customer information and the default path to a backup server, and (D) nightly incremental backups of the default path to a backup server; (ii) replicate Customer's database and default path to an off-site location (i.e., other than the primary data center); and (iii) save the last fourteen nightly database backups on a secure transfer server (i.e., at any given time, the last fourteen nightly database backups will be on the secure transfer server) from which Customer may retrieve the database backups at any time.

Use of Customer Information, Data Conversion, and Transition

Because the vendor will have access to and will be storing the customer's data, the agreement should contain specific language regarding the vendor's obligations to maintain the confidentiality of such information and confirming that the vendor has no right to use such information except in connection with its performance under the cloud computing agreement. Moreover, data conversion, both at the onset and termination of the cloud computing agreement, must be addressed to avoid hidden costs and being locked in to the vendor's solution.

First, it is becoming increasingly common for cloud computing vendors to want to analyze and use the customer data that resides on their servers for their own commercial benefit, in particular the data customers create as they use the software. For example, the vendor may wish to use a customer's data, aggregated along with other customer's data, to provide data analysis to industry groups or marketers. The vendor may limit its use to deidentified customer data, but this is not always the case. These uses are very similar to what businesses and individuals have been facing for years with respect to the ability of "cookies" to track and follow where we go and what we do on the Internet.

Here, however, the customer data in the cloud is proprietary and confidential to the customer and its business. As a result, the customer should consider such use of any of its data very carefully, and if the agreement does not mention these sort of uses, the customer should ask the vendor about its uses and add a vendor representation about which uses, if any, are permitted. It is commonly the case that customers conclude that the vendor should not have any right to use customer data, whether in raw form, aggregated, or deidentified, beyond what is strictly necessary to provide the software or for the vendor to perform its other obligations under the agreement. An example where commercial use might be acceptable is where the vendor provides a service that directly depends on the ancillary use of such customer data, such as aggregating customer data to provide data trending and analysis to the customer and similarly situated customers within an industry. In this case, it would be appropriate to specifically draft what customer information the vendor is permitted to access and what are the specific permitted uses.

Second, the customer must address data conversion issues and return of data upon termination of the cloud computing agreement and the relationship. Going into the relationship, the customer should know that its data can be directly imported into the vendor's software or that any data conversion needed will be done at vendor's cost or at customer's cost. It is commonly appropriate for a customer to conduct a test run of a vendor's mapping scheme to see how easy or complicated it will be (likewise when checking vendor's references, to ask about data migration experiences).

Lastly, the customer does not want to be trapped into staying with a vendor because of data formatting issues. To that point, cloud computing agreements commonly contain explicit obligations on the part of the vendor to return the customer's data, both in vendor's data format and in a platform-agnostic format, and thereafter destroy all of the customer's information on vendor's servers, all upon termination of the agreement. A sample provision to illustrate this obligation:

> At Customer's request, Vendor will provide a copy of Customer information to Customer in an ASCII comma-delimited format on a CD-ROM or DVD-ROM. Upon expiration of this Agreement or termination of this Agreement for any reason, (i) Vendor shall (a) deliver to Customer, at no cost to Customer, a current copy of all of the Customer information in the form in use as of the date of such expiration or termination and (b) completely destroy or erase all other copies of the Customer Information in Vendor's or its agents' or subcontractors' possession in any form, including but not limited to electronic, hard copy or other memory device. At Customer's request, Vendor shall have its officers certify in writing that it has so destroyed or erased all copies of the Customer information and that it shall not make any use of the Customer information.

Insurance

The customer should always address insurance issues in cloud computing situations, both as to the customer's own insurance policies and the vendor's insurance. Most data privacy and security laws will hold the customer liable for a security breach whether it was the customer's fault or the vendor's fault. Thus, the customer should help self-insure against cloud computing risks, including data and security issues, by obtaining a cyber liability or similar policy.

Cyber liability insurance can protect the customer against a wide range of losses. Most cyber insurance policies will cover damages arising from unauthorized access to a computer system, theft or destruction of data, hacker attacks, denial of service attacks, and malicious code. Some policies also cover privacy risks like security breaches of personal information, may apply to violations of state and federal privacy regulations, and may provide reimbursement for expenses related to the resulting legal and public relations expenses.

Requiring the vendor to carry certain types of insurance enhances the likelihood that the vendor can meet its obligations and provides direct protection for the customer. In addition to a cyber liability policy, other forms of liability insurance that a customer should require a vendor to carry in a cloud computing transaction include technology errors and omissions liability insurance and commercial blanket bond, including electronic and computer crime or unauthorized computer access insurance. These types of insurance will cover damages that the customer or others may suffer as a result of the vendor's professional negligence and by intentional acts by others (e.g., vendor's employees, hackers). It is critical that the customer require the vendor to have these sorts of policies and not just a general liability policy. Many commercial general liability policies contain a professional services exclusion that precludes coverage for liability arising from IT (e.g., cloud) services, as well as other exclusions and limitations that make them largely inapplicable to IT risks.

Indemnification

It is appropriate for cloud computing transactions to include a vendor indemnification whereby the vendor agrees to defend, indemnify, and hold harmless the customer, as well as the customer's affiliates and agents, from any claim arising out of the vendor's breach of its obligations with respect to the confidentiality and security of the customer's data. Any intentional breach should be fully indemnified, meaning that the customer will have no "out-of-pocket" costs or expenses related to recovery of the data and compliance with any applicable notice provisions or other obligations required by data privacy laws. In the event the data breach is not intentional, the vendor may require a cap on its potential liability exposure, which may be reasonable depending on the nature of the unintentional act and the type of customer data in question.

It is also appropriate for these transactions to include a broad intellectual property infringement indemnification that would protect the customer from damages, costs, and expenses arising out of any claim that the software infringes the intellectual property rights (think trademark, copyright, trade secret, patent, and any other intellectual property rights) of any third party. This means that the customer will never be responsible for any costs or expenses if some third party claims that the software the customer is using infringes its intellectual property. It is common for vendors to limit the intellectual property indemnification only to infringement of copyrights. This is not an appropriate or widely accepted limitation on the vendor's indemnity obligations, since many infringement actions arise out of patent or trade secret rights.

Vendors will also try to limit their exposure by limiting the indemnity to infringements of patents "issued as of the effective date" of the agreement. This limitation should be avoided since it will result in a customer's exposure to damages, costs, and expenses as a result of a claim that the software infringes a patent issued after the effective date. The vendor should be responsible for continued diligence with respect to the noninfringement of its software and any such limitation on that responsibility should be avoided. Vendors frequently also limit these indemnification obligations to "United States" intellectual property rights. While this is generally acceptable, the customer should consider whether its use of the software will occur overseas.

Limitation of Liability

The vendor's limitation of liability is very important in a cloud computing engagement because virtually all aspects of data security are controlled by the vendor. Thus, the vendor should not be allowed to use a limitation of liability clause to unduly limit its exposure. Instead, a fair limitation of liability clause must balance the vendor's concern about unlimited damages with the customer's right to have reasonable recourse in the event of a data breach, security incident, or other damages.

A vendor's limitation of liability clause usually (i) limits any liability of vendor to the customer to the amount of fees paid under the agreement or a portion of the agreement (e.g., fees paid for the portion of the software or services that gave rise to the claim), and (ii) excludes incidental, consequential (e.g., lost revenues, lost profits), exemplary, punitive, and other indirect damages. While a customer may not be able to eliminate the limitation of liability in its entirety, the customer should ask for the concessions in the following sections.

The Limitation of Liability Should Apply to Both Parties

The customer should at least be entitled to the same protections from damages that the vendor seeks.

- The following should be excluded from all (i.e., any exclusion of consequential damages as well as any cap on direct damages) limitations of liability and damages: (i) damages, costs, and expenses arising out of breach of the confidentiality and data security provisions by either party; (ii) claims for which vendor is insured; (iii) damages arising out of and costs and expenses to be paid pursuant to the parties' respective indemnification obligations; (iv) damages, costs, and expenses arising out of either party's infringement of the other party's intellectual property rights; and (v) damages, costs, and expenses arising out of breach of the advertising/publicity provision.
- The overall liability cap (usually limited to fees paid) should be increased to some multiple of all fees paid (e.g., two to four times the total fees or the fees paid in the twelve months prior to the claim arising). Keep in mind that the overall liability cap should not apply to the exclusions in the bullet point above.

License/Access Grant and Fees

The license or access grant in a cloud computing agreement encompasses three main issues: permitted use, permitted users, and fees. The grant as to permitted use should be straightforward and broadly worded to allow the customer full use of the software. For example, "Vendor hereby grants Customer a worldwide, nonexclusive right and license to access and use the software for Customer's business purposes." Vendor agreements often try to limit the customer's use of the software to "its internal purposes only." Such a restriction is likely too narrow to encompass all customer's desired uses. Drafting the license in terms of permitting the customer to use the software for "its business purposes" is a better, more encompassing approach.

The license rights related to which of customer's constituents can use the software, and at what price, can be far more complicated. As to permitted users, the customer must carefully define this in light of its needs and its structure. For example, beyond customer's employees, the customer may want affiliates, subsidiaries (now or hereafter existing), corporate parents, third parties such as outsourcers, consultants, and independent contractors all to have access to the software. The agreement should clearly set forth those users that fit the customer's anticipated needs.

There are many options with respect to pricing. A vendor may make software available on an enterprise basis, per user, per account, per property, per a specified number of increments of use or processing power, or per a specified number of megabytes of storage. They will often charge for storage in excess of a base amount. The customer's future use of the software is also an important consideration when negotiating fees. When entering into cloud computing agreements, anticipate and provide for the ability to add or remove users (or whatever unit the metric is based on), with a corresponding adjustment of the license fees. The best time for the customer to negotiate rates for additional use is prior to signing the agreement. Customers should also attempt to lock in any recurring license fees for a period of time (one to three years) and thereafter an escalator based on the consumer price index (CPI) or other third-party index.

Term

Because software delivered in a cloud environment is provided as service, like any service, the customer should be able to terminate the agreement at any time without penalty and upon reasonable notice (a few days to up to thirty days). The vendor may request a minimum commitment period from the customer to recoup the vendor's "investment" in securing the customer as a customer (i.e., sales expenses and related costs); however, these should be minimal since there is not a significant amount of upfront costs incurred by providers of cloud-based software applications. If the customer agrees to this, the committed term should be no more than a few months or, in some circumstances, up to one year, and the vendor should provide evidence that it has actually incurred the upfront costs to justify such a requirement.

Warranties

In a cloud computing agreement, the key warranties as to uptime and access are covered by the service levels. Warranties regarding data security, redundancy, and use were previously covered in this chapter. Beyond those critical warranties and service levels, several other warranties are appropriate to include in cloud computing agreements.

The vendor should warrant the following:

- The software will perform in accordance with the vendor's documentation (and any agreed-upon customer specifications).
- All services will be provided in a timely, workmanlike manner, in compliance with industry best practices.
- The vendor will provide adequate training, as needed, to customer on the use of the software.
- The software will comply with all federal, state, and local laws, rules, and regulations.
- The customer's data and information will not be shared with or disclosed in any manner to any third party by vendor without first obtaining the express written consent of customer.
- The software will not infringe the intellectual property rights of any third party.
- The software will be free from viruses and other destructive programs.
- There is no pending litigation involving vendor that may impair or interfere with customer's right to use the software.
- The vendor has sufficient authority to enter into the agreement and grant the rights provided in the agreement to customer.

Publicity and Use of the Customer Trademarks

The customer's reputation and goodwill are substantial and important assets. This reputation and goodwill are often symbolized and recognized through the customer's name and other trademarks. Accordingly, every cloud computing agreement (and most other IT agreements) should contain a provision relating to any announcements and publicity in connection with the transaction. The vendor should be prohibited from making any media releases or other public announcements relating to the agreement or otherwise using the customer's name and trademarks without the customer's prior written consent.

Notification for Security Issues

The cloud computing agreement should require that if a breach of security or confidentiality occurs, and it requires notification to customer's customers or employees under any privacy law (federal, state, or otherwise), then the customer must have sole control over the timing, content, and method of such notification. The vendor should be prohibited from notifying affected customer's customer unless specifically instructed by the customer to do so. These agreements commonly also contain a requirement that if the vendor is responsible for the breach (whether partially or fully), then the vendor must reimburse the customer for the customer's out-of-pocket costs and expenses associated with customer providing the notification—even if the customer was not required by applicable law to provide the notification but did so as a gesture of goodwill toward its affected customers or to preserve its reputation among its customer base.

Assignment

A provision prohibiting the customer from assigning its rights under the agreement should be avoided. Cloud computing agreements often permit the customer to assign its rights to its affiliates and other entities that may become successor or affiliates due to a reorganization, consolidation, divestiture, or the like. Any concerns the vendor may have from an assignment can be addressed by the requirement that the assignee will accept all of the customer's obligations under the agreement. Similarly, the customer should also obtain assurance that any vendor assignee will agree to be bound by all of the terms and conditions of the agreement, including without limitation service level obligations.

Pre-Agreement Vendor Due Diligence

Lastly, consider doing pre-agreement diligence on the vendor. By crafting and using a vendor questionnaire, the customer can, at the outset, get a good idea of the extent to which the vendor can meet the customer's expectations and business requirements. The customer can then identify where gaps exist so that they can be eliminated or so that the risks associated with the gaps can be reduced through negotiation of the vendor's requirements. Examples of the items to cover in such a due diligence questionnaire include the vendor's financial condition, insurance, existing service levels, capacity, physical and digital security, disaster recovery, business continuity, redundancy, and the ability to comply with applicable regulations.

Summary

In conclusion, cloud computing agreements, like traditional software license agreements, should be negotiated with the customer's needs in mind as vendor forms are invariably one-sided. Unlike traditional software licenses, the customer should focus less on configuration of the application and more on its availability and the security of its data.

Chapter 7

Click-Wrap, Shrink-Wrap, and Web-Wrap Agreements

CHECKLIST

Where's the Agreement?

- ☐ Identify all relevant contract terms
- ☐ Keep accurate copies of the agreements
- ☐ Record the date of acceptance

Risks and Issues

- ☐ Business assessment of risks posed
- ☐ Understand as-is nature of software or service
  - No warranties
  - No indemnities
  - Very limited liability, if any, for the vendor
- ☐ Customer has unlimited liability for both direct and consequential damages

- ☐ Identify contractual provisions that could place customer intellectual property at risk
  - Feedback clauses
  - Train personnel not to reveal or disclose proprietary information or intellectual property
- ☐ Avoid placing sensitive information at risk
  - No confidentiality protection
  - No real liability for breaching confidentiality
  - No remedy for misuse of confidential information
- ☐ Beware of broad audit rights
  - Abusive audits
  - Contingent fee audits
  - Access to company facilities and systems without adequate contractual protections
  - Access by undefined third-party agents of vendor
- ☐ Assess risks of use of resellers
  - Additional contract to review
  - Splitting of responsibility
  - Potential finger pointing between vendor and reseller

Techniques

- ☐ Blind acceptance
- ☐ Knowing acceptance
  - Identify all terms
  - Conduct brief review and assessment
  - Adequately document
- ☐ Mitigation
  - Process for review
  - Development of form amendment

Overview

Click-wrap, shrink-wrap, and web-wrap agreements are the fine print you see, among other things, when you click through terms and conditions in accessing an online service (e.g., in connection with a cloud computing service) or as part of the installation process for a piece of software. They may also be encountered as part of the documentation provided with new software or a hardware component. They may even be found, with some searching, in a file entitled "license.txt" or similar name on the installation CD on which a new piece of software is delivered.

Companies seldom read these terms in any detail, generally view them as nonnegotiable, and accept them as a necessary evil.

The fact is, these types of agreements can present significant legal and company issues. They can place a company's sensitive data at risk, expose the company to liability, compromise the company's ownership of its own intellectual property, and cause the company to pay additional, unforeseen fees.

When the first edition of this book was written, these types of shrink-wrap/click-wrap agreements were generally found only in small engagements, involving off-the-shelf software. Since publication of the first edition, however, we have seen these types of agreements used in fairly substantial, more business-critical engagements. Yet, the vendor presents the agreement as "non-negotiable." As described below, there are ways to mitigate this risk, but as the size and criticality of the engagement increase, the customer should explore alternate vendors with more appropriate contracting practices.

What Is a "Shrink-Wrap" License?

The term "shrink-wrap" derives from the method by which software was distributed as a package of installation disks and associated documentation sealed inside shrink-wrap cellophane. The accompanying end-user license agreement was often itself packaged in shrink-wrap cellophane and placed on the outside of the package or included as the topmost item in the package. Today, shrink-wrap agreements can take a variety of forms and are found in both software and hardware acquisitions. However, they all have a common structure: essentially nonnegotiable terms and conditions that accompany the product. The terms may appear as part of the documentation accompanying the product, as part of an online purchase process whereby the terms are displayed (and the purchaser, potentially, is required to affirmatively click an "accept" button as part of the process), or presented to the purchaser on first use of the application as part of the installation process.

If the terms are displayed electronically, either online or in connection with the installation process, they are often referred to as "click-wrap" terms. For purposes of this discussion, there is no difference between click-wrap and shrink-wrap terms.

Courts in the United States have almost uniformly found that these types of agreements are enforceable (see, e.g., *Conference America Inc. v. Conexant Sys. Inc.*, M.D. Ala., No. 2:05-cv-01088, 9/10/07). In fact, courts have held them enforceable even if the customer failed to read them (e.g., *Druyan v. Jagger*, S.D.N.Y., No. 06-cv-13729, 8/29/07). Except in unusual situations involving a very narrow range of unique transactions, these types of agreements have almost uniformly been found to be enforceable in the United States.

Products Purchased under Shrink-Wrap Agreements—Common Elements

While there are no bright-line rules as to the specific types of products that are made available under shrink-wrap agreements, the following are common elements:

- The product typically has a relatively low cost per unit (e.g., less than $20,000). While the cost per unit for a given product may be low, or even trivial (e.g., less than $100), the total cost to the organization should not be overlooked (e.g., 1,000 units at $100 per unit results in aggregate fees of $100,000). An easy example would be a copy of Microsoft Word or Adobe Acrobat. Essentially all open-source software is licensed under shrink-wrap terms.
- The product is provided "off-the-shelf," meaning that it is not customized for the purchaser. Each purchaser purchases the exact same version of the product as every other purchaser, without modification.
- The product requires very little implementation effort. The purchaser generally assumes all of the installation effort without obtaining professional services from the vendor or a third party.
- The product is generally not mission-critical.
- The product is typically well understood and established in the marketplace. Frequently, the product is available for trial and evaluation before a license is required.

The listed points are, of course, only generalities. It is important to note that there are many instances in which shrink-wrap agreements are used for the purchase of products that cost hundreds of thousands of dollars, require extensive customization and a significant implementation effort, and are mission-critical to the organization. As discussed in the next section, the risk of the products purchased under a shrink-wrap model can increase dramatically when the proposed application varies from the foregoing common elements.

Methods of Purchasing Shrink-Wrap Products

There are essentially two means of purchasing shrink-wrap products. First, the product can be directly purchased from the vendor that created it (e.g., downloading a copy of Adobe Acrobat Reader from Adobe's website or purchasing a copy of Word directly from Microsoft). Second, the product can be purchased through a reseller or similar entity that is authorized by the vendor to distribute the product.

One benefit of using a reseller is the potential to license and purchase products, particularly large orders, at a substantial discount. Another advantage is the possibility of negotiating an enterprise or master contract with favorable legal and business terms for all licenses and purchases made through the reseller. In many instances, however, the use of resellers results in the licensee or purchaser obtaining substantially less favorable terms than if the licensee or purchaser directly negotiated with the vendor and eliminated the use of the reseller. Resellers generally insist on highly protective agreements that absolve them of liability for the products they distribute. Any protections relating to the products are provided in the form of nonnegotiable shrink-wrap agreements from the manufacturers or, worse yet, provided through websites that may change at any time. In either case, the product terms are (i) nonnegotiable, and (ii) almost always very minimal, offering little in the way of substantive warranties and indemnities. A growing number of manufacturers are turning to reseller arrangements for the express purpose of avoiding having to extend appropriate, market-based contractual protections to their customers.

Reseller arrangements should generally only be considered when the product satisfies the common elements described above (e.g., low fees, noncritical use, off-shelf, well established, potentially trialed) and the cost–benefit of proceeding with transaction is justified. This usually means the reseller will be used for the purchase of a narrow range of preapproved products for the organization, like purchases of standard office productivity applications (e.g., Microsoft Word and Adobe products).

Typical Shrink-Wrap Terms and Conditions

While the type of terms and conditions found in shrink-wrap agreements varies greatly from vendor to vendor, there are a number of common themes. In general, shrink-wrap agreements include the following potentially problematic terms:

- Little or no warranty protection. In most instances, all warranties are expressly disclaimed—meaning the software is provided entirely "as-is."
- There is generally no protection in the event the purchaser is sued for intellectual property infringement arising out of its licensed use of the products. For example, a purchaser could be sued for patent infringement arising out of use of a software product and—even though the vendor is the cause of the infringement because of the way it developed the software—find itself with no protection under its software license agreement with the vendor. These types of claims have become more and more prevalent. In fact, entire businesses have been founded based on developing large patent portfolios and then, as their revenue source, suing the licensees of software for damages. Most negotiated agreements include an indemnification from infringement claims.
- A limitation of liability that absolves the vendor of all or substantially all liability for all damages of every kind and type. If an indemnity for intellectual property infringement is provided, the indemnity is generally subject to

the overarching limitation of liability, significantly diminishing the vendor's obligation to indemnify.

- In contrast, the purchaser will have unlimited liability for all forms of damages. The purchaser may also be required to give the vendor a broad and frequently poorly defined indemnity for a wide range of claims, some of which may arise from the vendor's own conduct.
- Little or no protection for confidentiality of the purchaser's information. The lack of this protection is a critical risk if the vendor has the right to access the purchaser's facilities and systems to conduct audits. Shrink-wrap and click-wrap agreements frequently contain specific language permitting the vendor to have broad rights to conduct onsite audits of its customers' facilities and computer systems, frequently with little or no notice. Those audits can expose highly sensitive information of the purchaser.
- The location (venue) at which a potential litigation or arbitration must be conducted may be inconvenient for the purchaser. For example, a purchaser in California may be required to arbitrate a dispute under the agreement in Florida. If the value of license is only, say, $10,000, having to engage an attorney and attend meetings in Florida would be cost-prohibitive.

These are general observations only. The specific language of a given shrink-wrap agreement may present additional risks. In particular, as discussed in the next section, a growing number of shrink-wrap agreements may present substantial risks to the purchaser's own intellectual property or, if the purchaser is in a regulated industry (e.g., financial services or healthcare), to the purchaser's data.

Key Risks of Shrink-Wrap Products

This section presents some of the key considerations and risks presented by shrink-wrap, web-wrap, and click-wrap agreements. In the following section, various risk mitigation strategies are discussed.

- **Where's the agreement?** In many cases, just locating the applicable agreement(s) may present a challenge. It is not uncommon for these types of agreements to reference other agreements, including other terms provided on other web pages. In addition, as mentioned previously, in some cases the agreements are hidden in licensing files on the installation disk for the software. Finally, once presented to the user, it may be all but impossible to later find the agreement (or version of the agreement) actually accepted by the user.
  - When presented with these types of agreements, customers should make sure to print or otherwise retain copies of the agreement, including the date on which it was accepted.

- **No remedies.** The end result of the terms and conditions commonly found in shrink-wrap agreements, as discussed in the preceding section, is that the purchaser has little or no remedy against the vendor in the event there is an issue with the product or if damages arise (e.g., the product has a substantial bug in it, ceases to function, causes an intellectual property infringement claim) out of use of the product. The product is, essentially, being licensed on an "as-is" basis. In most instances, the purchaser's only remedy in the event of a problem is to cease use of the offending product. A refund or other compensation is unlikely.
- **Safety in numbers.** In general, the purchaser's primary protection in purchasing shrink-wrap products is the concept of "safety in numbers." That is, the product is widely distributed and usually well established in the community. This reduces the potential for a substantial bug or defect to go without a fix from the vendor. The purchaser is essentially relying on the power of the market to force the vendor to correct issues (i.e., vendors with poorly designed or buggy products will lose market share and, at least arguably, be easy to identify).
- **Risks to customer intellectual property.** Some shrink-wrap agreements contain expansive "feedback" and similar clauses that could result in the licensor gaining ownership of the purchaser's own intellectual property. The contract actually includes language that the purchaser is assigning its intellectual property rights to the vendor. In some cases, almost anything the purchaser shares with the vendor, including during support discussions, may become the vendor's property or, at minimum, result in the vendor having an unbridled license to use what it has learned for its own business purposes. At best, this can result in the purchaser essentially granting the vendor a free license to the purchaser's valuable intellectual property. At worst, it can result in the purchaser losing all control over its intellectual property.
  - The only way to control this risk, absent declining to accept the contract, is to carefully coach all personnel having contact with the vendor not to reveal or discuss proprietary information and intellectual property of the customer.
- **Beware of broad audit rights.** Shrink-wrap agreements may also include broad audit rights, permitting the vendor almost unlimited access to the purchaser's facilities, records, and systems. In some instances, these rights permit any or all of the vendor's agents, contractors, and licensors to also have full access to the purchaser's facilities, records, and systems. Under these terms, purchasers assume the additional risk of having third parties, with whom the licensee has no contract and no confidentiality protection, unfettered access the licensee's facilities, records, and systems. For regulated entities (e.g., in financial services and healthcare) and all others in possession of consumer information, these audit rights subject the licensee to the additional risk and

potential of exposing highly sensitive and regulated data to vendors and other third parties without adequate contractual protections (e.g., confidentiality clauses, information security protections, and limitations on use). Consider the potential risk presented by a vendor showing up at a purchaser's facility, without notice, and demanding full access to its systems and records—without any protection for the purchaser's highly sensitive confidential information and data or any protection if that access causes a disruption in the purchaser's operations.

- **Abusive audits**. Audits can also be excessive and abusive, disrupting the licensee's normal operations and potentially making the licensee liable for substantial financial liability for third-party auditor fees (which can reach the hundreds of thousands of dollars). This is because many vendors view these audit rights as a means to derive additional revenue from their purchasers. Some auditors even work on a contingency basis, forcing them to either find a problem or forego payment. This creates an undue incentive for the auditor to search until it finds something. In a number of instances, audits have led to substantial additional fees being paid by purchasers in agreements that were not properly negotiated. In one case, an audit revealed a relatively minimal excess use of the software, which resulted in the payment of a few thousand dollars in additional license fees. Unfortunately, the customer was also responsible for paying nearly $40,000 in audit costs. Given the current economic climate, vendors are conducting these audits on an ever-increasing basis to try to squeeze more revenue from their customers. The headlines are full of instances in which companies have paid substantial additional fees for excess license uses.
- **Avoid placing sensitive information at risk**. Given the as-is nature of the software or service and the lack of any substantive contractual protections, customers should generally avoid placing any highly sensitive information at risk in connection with the engagement (e.g., refraining from uploading confidential personally identifiable information of consumers to a web-based service under a click-wrap agreement that affords no real protection for that information).
- **Reseller issues**. With regard to reseller relationships, additional risk can arise in situations in which the reseller is providing support or subcontracted support for the licensed product. Splitting the agreements governing the purchase of the product from support obligations and having two different responsible contracting parties can lead to finger pointing when failures occur and leave a customer without adequate remedies to bridge the two agreements (e.g., if the purchaser purchases a piece of hardware and the reseller breaches its support agreement, the customer may be able to show damages under the support agreement, but will likely have no claim or remedy under the purchase agreement).

Mitigating Risk

There are essentially three methods of addressing the risk of shrink-wrap agreements: blind acceptance, knowing acceptance, and mitigation.

- **Blind acceptance.** Blind acceptance refers to the practice of looking at a proposed use of a product, ensuring it falls within the common elements of shrink-wrap products identified above (e.g., low fees, noncritical use, off-shelf, well established, potentially trialed), and electing to proceed with the purchase without further consideration. Few sophisticated organizations take this approach. It would require the purchaser to proceed without regard for the risk—abandoning any effort at due diligence.
- **Knowing acceptance.** Knowing acceptance refers to the process of quickly reviewing the applicable license agreement for a proposed purchase of a shrink-wrap product and assessing whether it presents any unique risks (i.e., something beyond the typical terms identified above). Unless a unique risk is identified or the purchase would present conditions beyond the common elements identified above, the transaction is approved. If unusual or unique risks are present (e.g., the aggregate value of the transaction is substantial, or the contract presents risks to the purchaser's intellectual property or data), the risks would be clearly identified in a memorandum for review and—if the cost–benefit of the engagement warrants—potential approval by senior management. This is the most prevalent means employed by sophisticated organizations to address risk in transactions of this kind.
- **Mitigation.** The mitigation approach is used in circumstances where the relevant license agreement presents unusual risks or in situations where the purchaser operates in a regulated industry where the protection of data and contracting requirements, in general, are of heightened concern. It has become common in those industries to review proposed uses of shrink-wrap products as they would for any other product purchase transaction. With due regard for the relatively limited ability of purchasers to negotiate these types of agreements, purchasers quickly assess the risks posed by a new engagement and focus on mitigating only the most substantial risks. This is commonly done in the form of an amendment to the shrink-wrap agreement. Such amendments are usually brief, addressing only terms like basic warranties, basic infringement indemnity, audit rights, and protection of the purchaser's own intellectual property. A number of large organizations are now using these types of amendments to quickly mitigate key risks in these engagements. Their acceptance by vendors, particularly in larger transactions, is growing. If the amendment is rejected by the vendor and no alternate vendor of a similar product is

readily available, the risks would be clearly identified in a memorandum for review and, if the cost–benefit of the engagement warrants, potential approval by senior management.

The mitigation approach presents the most mature approach to addressing risk in shrink-wrap engagements.

Summary

Except in rare instances, shrink-wrap, web-wrap, and click-wrap agreements are enforceable. As with any contract, they must be reviewed and assessed to identify risk. The business can then conduct a cost–benefit analysis to determine whether the risk is warranted and whether that risk can be controlled, at least to some degree, through the use of the mitigation approach discussed in this chapter. The risks presented by shrink-wrap, web-wrap, and click-wrap agreements should not be minimized.

Chapter 8

Maintenance and Support Agreements

CHECKLIST

Fee Predictability

- ☐ Price locked for a fixed term
- ☐ After rate of price increases capped at CPI

Term

- ☐ Minimum period of support

Termination and Resumption of Support

- ☐ Ability to resume support after earlier expiration or termination specifications
- ☐ Support obligations tied to "specifications" rather than "documentation"

Availability

- ☐ Support available 24/7 if needed

Problem Escalation

- ☐ Support escalation matrix

Service Levels

- ☐ Escalation matrix response time service level
- ☐ Service credits for service level failures

Limitations of Liability

Overview

Paid maintenance and support services provide protection following the warranty period. "Support" includes technical support services such as documentation, telephone help desk support, on-site support, and error correction/bug fixes when the software or service doesn't work in accordance with the specifications. "Maintenance" would include the vendor's obligations with respect to keeping the software or service current via updates, upgrades, enhancements, new releases, and the like. The following are some key considerations for any maintenance and support agreement.

In general, the vast majority of support and maintenance terms are written using aspirational language (e.g., "response time goal," "targeted resolution time"), as opposed to firm contractual commitments to resolve issues within defined periods of time. As such, the support and maintenance terms are no substitute for well-drafted acceptance and warranty provisions.

Scope of Support and Maintenance

- It is important that any agreement clearly describes the support that will be provided by identifying what software and services will be supported and the amount and type of support that will be provided. Before an agreement has been signed, companies should also inquire as to which support services are *not* included and will only be provided for an additional fee, so that such information can be factored into the negotiation process. Although vendors often argue that all of their clients receive the same maintenance and support, companies should not be afraid to insist on the level of support that is necessary to support their business needs during negotiations.
- Requirements should be specified in detail for when and how updates can occur to avoid interference with the company's use of the software or service.
- Vendors should provide detailed documentation and make training available to customer's personnel who will interact with the support software or service.
- It has become common for vendors to limit the amount or type of support services included without additional charge. These arrangements can introduce significant uncertainty for companies regarding what fees they will be charged in the event of a problem with the software or service. This

uncertainty can create a reluctance to utilize support services, which exacerbates the problem with the software or services and delays its resolution. The company should seek to reach an arrangement with the vendor where all support activities are included in a flat fee. While vendors typically resist requests for such "all included" provisions, this approach is often worth the extra effort during negotiations, especially for business-critical software or services where going without support and maintenance services is not a viable option.

- If a vendor rejects an "all included" approach, companies must be certain that they understand the range of possible additional charges they may face over the life of the agreement so that they can factor those additional support costs into their vendor selection process. These arrangements are a lucrative source of revenue for vendors. By enabling them to charge for minimal support, and then charging extra for new releases, fixes to their defective product necessary technical support, reasonable response times, and on-site support to fix their defective products, vendors can deflate upfront costs in a way that makes it difficult to make comparisons with other vendors. Before entering into an agreement, a company should require the vendor to provide a detailed listing of all support services that are not included into the pricing structure.
- However, most vendors will generally reject company requests to rewrite or expand the company's definitions of bug fixes and updates, so they can avoid giving new versions or expanded functionality, and thus require additional fees.

Predictability of Fees

- Maintenance and support fees are typically a percentage of the actual license fees that are paid by the company to the vendor. These can vary widely, but are commonly in the 15%–20% range. As discussed previously, uncertainty regarding additional fees can complicate the support relationship. Likewise, uncertainty over what support fees will be charged after the initial term can make it difficult to determine a software or service's total cost of ownership and make comparisons between the solutions offered by different vendors.
- Companies should consider locking pricing on support and maintenance for a fixed period of time (e.g., three years) and then negotiating a cap on maintenance and support fee increases over time (e.g., fees may increase no more than the percentage change in the applicable consumer price index (CPI) or 2%).
- If a vendor resists such a formula, other alternate indices may be more appropriate, including industry-specific indices produced by the Department of Labor, or a CPI that excludes one or more components from a more standardized index. A company may wish to address the adjustment to an agreed-upon percentage for particular fees separately where its pricing is influenced by either unique factors such as actual costs of labor or other components or the return on initial investment for technology equipment.

- Some vendors provide companies with a "bank of hours" for support after which additional hours are charged at a set fee. The company may be faced with substantial fees under this approach in the event of a serious problem. One solution is to negotiate a "stair-step" approach where a larger number of hours are provided during implementation and for several months thereafter, when support calls are likely to be high. The number of hours can be reduced thereafter. Warranty-related calls and product defect calls should not be deducted from a customer's bank of hours.
- As discussed in Chapter 24:
  - Software is generally heavily discounted off of vendors' list prices, so it is important for companies to ensure that their support pricing reflects the appropriate percentage of the discount price actually paid for the software, not the list price.
  - Vendors will generally resist attempts to fix time and materials rates for professional services beyond one year.
  - Many vendors will insist on the ability to charge for their time when a company's support problem turns out to be caused by the company's own products or other external factor.

Support Not to Be Withheld

- Except for the failure to pay undisputed amounts for support, there should be no basis for a vendor to withhold a company's support and maintenance services. Any other action or inaction by the company can be addressed by a vendor-initiated claim for damages. Ongoing maintenance and support are necessary for optimal use, especially where the software or service performs critical business functions. Companies should address this issue early by negotiating language that assures that the company will not be "held hostage" by the vendor in disputes. For example,

  Support under this agreement will not be withheld due to any unrelated dispute arising under this agreement, another agreement between the parties, or any other unrelated dispute between the parties.

Term

- Obtaining a multiyear "initial support term" with a series of optional renewal terms is generally the best approach to provide a company with stable support with predicable fees. This approach commits the vendor to providing support and maintenance for the company's software or service over the entire support term (both initial and extended support terms), while only obligating the company for the initial support term. It also provides a basis for fixing prices for the initial support term and obtaining agreed-upon escalations of maintenance and support prices during the extended support terms. While the length of initial support terms can vary widely depending on the size of the transaction and the nature of the service or software, initial terms of three to five years in duration are common. Vendors typically offer between two and five optional extended support terms.

- Companies should be aware of whether their agreements contain automatic renewal ("evergreen") provisions. These provisions often contain limited windows of time at the end of the term during which the customer must notify the vendor of its intent not to renew the agreement. In the absence of such notification, agreements typically renew automatically for a renewal term of similar duration to the previous term. The company may attempt to affirmatively manage these nonrenewal notification periods by calendaring such dates at the time of executing the agreement. However, the extended length of these terms and turnover among company's personnel may cause these dates to pass unnoticed. To assist the company, language should be included in the agreement requiring the vendor to provide the company with prior written notice of the end of the initial support term and each extended support term. This notice should be given in advance of the required renewal notification period, to allow the company to evaluate the terms of the current agreement and marketplace alternatives, so it can make an informed choice of whether it would like to renew.

- Companies who are uncomfortable with automatic renewal can include language that requires that renewals are only exercised upon affirmative written notice from the customer of its intent to renew (i.e., options to extend are not automatically exercised). However, customers must be mindful to avoid missing the deadline for notice of renewal and having their agreement, and by extension, any hard-fought fixed pricing terms terminate automatically.

Partial Termination/Termination and Resumption of Support

- Maintenance and support services are necessary only to the extent the solutions they support are in use. A customer's business operations, including the demand for its products or services, the types of products or services it offers, and by extension, the services and software it uses in its business, should be expected to fluctuate over time. As such, it is important that the terms of any agreement allow for adjustments in the level of support to reflect changes in the use of the underlying software or service.

 - Language permitting a customer to partially terminate a portion of its agreement can provide the flexibility necessary for businesses to adapt to fast-changing market conditions. Customers should consider tying their ability to terminate a portion of the services in their Agreements to the termination of the underlying software or service itself.
 - Maintaining the customer's ability to terminate maintenance or support services and then resume such services within a specific period of time can contribute to cost savings by reducing the customer's reluctance to cancel currently unused maintenance or support services.

Specifications

- It is common for agreements to limit the defects of software or services that are covered under the agreement to those that result in a failure to "substantially perform" in accordance with their then-current published documentation. The claims made in a vendor's documentation are typically much more conservative than those made by the same vendor's promotional materials and personnel during the sales cycle. As a result, there may be a significant gap in the functionality a customer has been "sold" and the functionality that the vendor is capable of supporting. Additionally, as each vendor controls the contents of its documentation, a vendor is able to unilaterally revise its documentation under this approach, which may result in significant degradation of the software or service's supported functionality. Although vendors will resist such a change, customers are on strong ground to insist that significant investment they have made on a solution is supported to a broader standard. Customers should consider language that requires the software or service to perform in accordance with the "specifications." These specifications should reference an exhibit that lists important functionality as well the terms and conditions of the agreement with the vendor and, then to the extent it is not inconsistent, reference the vendor's documentation. For example,

  Specifications shall mean the specifications and requirements set forth in exhibit a (specifications/requirements), all other performance requirements included or incorporated by reference into this agreement, and, solely to the extent not inconsistent with the above, the documentation.

Availability

- Customers should carefully evaluate whether their business needs necessitate the availability of support services on a 24/7/365 basis. The availability of such support can make the difference between a prompt resolution of an issue and an extended disruption of service, along with its associated costs—particularly for business critical applications, even if this off-hours support is charged as an additional cost. If a vendor is unwilling or unable to provide such support for their products, that factor should be a significant consideration in the vendor selection process.

Support Escalation

- It is critical that a customer is familiar with a vendor's support escalation procedures before a problem arises. The vendor should provide the customer with a detailed description of the escalation of a problem from its first-tier support to second-tier support. The customer should also insist that the vendor provide, and continue to update, a support escalation matrix to be used in the event that a mission-critical disruption requires immediate escalation. The support escalation matrix should identify multiple individuals comprising the vendor's support management team, along with their contact information.

Service Levels

- The level of service a customer will require, such as for responses and resolutions to issues, is proportional to the criticality of the supported software or service. As vendors continue to offer solutions that serve in increasingly critical roles, they must also be willing to provide higher levels of service to correspond with their customers' exposure to risk. The following issues should be considered when negotiating the service level of agreements:
  - A maximum amount of time should be specified for the vendor to respond to a customer's support request. A "response" should require that the vendor has engaged on the support request; is working continuously to diagnose the corresponding errors, formulate a plan to address any such errors, and execute that plan; and has notified the customer that such support has begun.
  - It is also appropriate to specify required problem resolution times. Unless the vendor has independent notification of the problem, resolution times are typically measured from the time when the vendor receives the support request until the time the vendor has resolved the problem. Customers should consider negotiating language that defines a "resolution" as both the vendor providing a correction of the problem and the customer confirming that the vendor's action *actually* corrected the problem. This approach will prevent the vendor from "resolving" the support request before a problem has truly been resolved in order to avoid defaulting on its service level obligations.

- Vendors of services are often in the best position to learn of a failure or other disruption of service. Their service levels should include an obligation that requires prompt customer notification and automatic initiation of a support request in the event of a problem.

One frequently used technique to mitigate risk in support and maintenance agreements is to choose the highest (and, potentially, second highest) issue categories (e.g., Categories 1 and 2 or Priority 1 or 2) and attach a clear required cure period to those important issues. For example: "If Vendor fails to resolve a Priority 1 issue within fifteen days (and such time is not extended by the written agreement of Customer), Customer may terminate this Agreement on written notice to Vendor as a material, non-curable breach."

Summary

Maintenance and support services agreements are integral for providing protection following the warranty period. Customers should consider the scope of the agreement, the exact type of support it needs for its software and services, and how to draft an agreement that obligates vendors to provide the appropriate level of support.

Chapter 9

Service Level Agreements

CHECKLIST

Common Provisions in Terms and Conditions

- ☐ Root cause analysis
  - Identify reasons for failure
  - Develop corrective action plan
  - Implement preventative corrections
- ☐ Cost and efficiency reviews
- ☐ Continuous improvements to service levels
- ☐ Termination for failure to meet service levels
- ☐ Cooperation

Common Provisions in Service Level Agreements and Attachments

- ☐ Measurement window
- ☐ Reporting requirements
- ☐ Maximum monthly at-risk amount
- ☐ Performance credits
  - Specify amount or percentage
  - Total category allocation pools
  - Specify parameters, including timing and notice requirements
- ☐ Presumptive service levels
- ☐ Exceptions to service levels
- ☐ Supplier responsibilities
- ☐ Additions, deletions, and modifications to service levels

- ☐ "Earn-back" of performance credits
- ☐ Map the form of service levels
  - Title
  - Measurement window
  - Actual service level *or* expected and minimum service levels
  - Calculation for how service level is derived

Overview

Service levels are an essential tool in many different types of information technology (IT) contracts to ensure proper performance of the services and supplier obligations and user satisfaction. IT agreements, whether professional services agreements, software license agreements, reseller agreements, cloud computing agreements, or other types of IT agreements, should contain precise terms with respect to the obligations of the supplier to perform its obligations consistent with and, at a minimum, in accordance with the specific service levels that are articulated throughout the agreement and in a service level exhibit that is attached to and incorporated into the underlying agreement. It is important to note that provisions related to service levels are commonly included throughout an agreement and also specified in a service level exhibit.

As mentioned in the preface to this Second Edition, one of the most alarming trends in the industry is vendor reluctance, particularly in cloud engagements, to commit to meaningful service levels. Instead, they include service level language that looks like "real" service level protection, but is, in fact, largely illusory. On average, 30% of the service level language we review fails into this category. It would literally be impossible for the vendor to ever breach service level commitments. That is why, these same vendors proudly claim they have not had a service level failure in over a decade. It is an easy thing to say when there is no way the vendor could ever actually fail to achieve the requirements. They are simply so heavily qualified and diluted, that the customer receives almost no protection, including even a basic right to terminate the agreement. All too often, customers are tied to contracts paying substantial monthly fees for a service that is not performing with no means of getting out. It is for these reasons that customers should review service level language with the greatest care.

Service Level Provisions Commonly Found in the Terms and Conditions

The following types of service level provisions are commonly found in the body or terms and conditions of an IT agreement. These types of provisions contain terms specific to service level failures and improvements over time. Specific details with respect to the service levels are more commonly detailed in a separate service level exhibit to the agreement.

Root Cause Analysis, Corrective Action Plans, and Resolution

In the event that the supplier fails to provide the services in accordance with the service levels articulated in the service level exhibit, after a reasonable period of time after notice, the supplier should be obligated to perform a root cause analysis to identify the reasons for the failure and describe the process for developing a corrective action plan and implementing corrections to prevent the failure from occurring in the future. The agreement should also describe the supplier's obligations with respect to correcting failures (at no cost to the customer) within a specified period of time or, if corrections cannot reasonably be implemented within the time specified, then a requirement that the supplier provide reasonable assurances that corrective steps will be taken that will result in a permanent correction and, in the meantime, workarounds have been implemented to prevent future failures while a permanent correction is being developed. It is imperative that these corrections, workarounds, and steps not result in any increased fees payable by the customer to the supplier. Each corrective action plan must contain, at a minimum, the following information and requirements:

- A commitment by the supplier to devote the appropriate amount of time and supplier resources (such as skilled personnel, systems support, and equipment) to prevent further occurrences of the service level failure.
- A strategy for developing fixes and improvements to prevent any of the same service level failure from occurring in the future.
- A detailed project plan with timeframes for implementing the proposed and required corrective actions.

Where the root cause analysis reveals that the service level failure was not caused by the supplier, the agreement should require that the supplier continuously and timely cooperate with the customer to correct the failure. While the best case is that the supplier provides these services at no additional charge to the customer, it is reasonable and common for the supplier to charge reasonable fees for such services, which should not exceed the fees charged for similar services charged to the customer or supplier's other customers. Where the root cause analysis demonstrates that the customer and supplier both shared in the cause of the service level failure, these same obligations of supplier should apply, but should be at no cost to the customer.

Cost and Efficiency Reviews

As part of the service levels provided by the supplier to the customer, a common requirement is that the supplier provide cost and efficiency reviews of the services being provided and for the supplier to make recommendations to the customer for reducing the cost of the services. These recommendations can include methods to efficiently utilize resources chargeable to the customer including:

- tuning or optimizing the systems used to perform the services;
- using and analyzing the results of predictive modeling, trend analysis, and monitoring tools;
- analysis and isolation of application and infrastructure design, configuration, and implementation flaws;
- recommending aligning technology processes, tools, skills, and organizational changes with the customer's business requirements; and
- employing new technologies in use by the supplier to replace existing technologies used by the supplier to provide the services, even if the use of such new technologies will result in a reduction in monthly revenues to the supplier under the agreement.

In addition to the requirement to provide cost and efficiency reviews, suppliers are commonly required to include in their annual recommendations employment of new technologies (generally those that are made available by the supplier to its other customers for a specified period of time) to replace existing technologies used by the supplier to provide the services to the customer. If the customer can demonstrate that employment of such new technologies would result in a reduction of the fees for the services and if the customer elects to implement such new technology in accordance with the agreement, then the customer would commonly be entitled to a credit off of any fees associated with the implementation of such new technologies by the supplier.

Continuous Improvements to Service Levels

While the parties will agree to certain service levels when an agreement is signed, service levels tend to improve over time, and a mechanism should be included in the agreement to ensure that the service levels can be updated to reflect the levels of service that are being provided by the supplier. For example, if a supplier is providing a service availability service level of 98%, but consistently for six or nine months the service availability has been 98.9% or more, the customer has a strong argument that the service level should be increased to reflect the actual level of service that is being provided by the supplier, not the service level initially agreed to by the parties. The agreement should describe a process for implementing new service levels based on supplier performance. For example, the parties could agree that a service level measurement standard could be adjusted to the average of the highest nine months of actual service level results achieved in any twelve-month period if such average exceeds the then-current agreed-upon service level. While suppliers are commonly willing to implement such a process, suppliers will frequently want to cap service level increases. For example, with an availability service level, a supplier might implement a cap of a high percentage availability. Or, for an availability service level, the parties could agree to a formula such that, for example, no service level will be increased more than 20% of the difference between 100% and the current service level.

Whatever the parameters, it is important for the parties to recognize that in most transactions that involve services levels, certain service levels are not meant to remain stagnant over the life of the contract. Instead, the parties should work together and implement processes to modify the service levels over time to reflect the actual levels of performance being provided by the supplier.

Termination for Failure to Meet Service Levels

While one or a few service level failures over time will result in a performance credit or other remedy available under the agreement, repeated service level failures can have a significant impact on the customer, and the value of the services provided can so significantly decline that the services no longer support the customer's operation or have value to the customer. Perhaps, too, for the fees being paid by the customer, the diminished service being provided results in the customer paying inflated prices.

A termination right may arise if the supplier consistently fails to meet the service levels. For example, a termination right might be appropriate if the supplier fails to meet the same service level for three consecutive months or for three months in any six-month period. In agreements in which there are both *critical service levels* (which commonly have higher performance credits and remedy obligations and are typically reserved for the most important services being provided by the supplier) and *key performance indicators* (which often carry reduced or no performance credits and are typically reserved for more routine business processes that do not impact mission critical processes), termination rights for failure to meet the service levels may differ. For example, it might be appropriate for the customer to have a termination right if the supplier fails to meet a critical service level for three consecutive months, whereas it might be appropriate for a customer to have a termination right if the supplier fails to meet a key performance indicator for six consecutive months. Due to the difference in severity and impact resulting from missed critical service level and key performance indicators, the termination rights are likely to be more aggressive with critical service levels.

Cooperation

In many cases, the supplier's ability to provide the services in accordance with the service levels, and to promptly respond to and resolve service level issues, requires that the supplier work not only with the customer, but also with the customer's other suppliers, third-party vendors, subcontractors, and other people and entities. Cooperation between the supplier and all of these stakeholders is essential and should be a requirement in the agreement. Suppliers are frequently obligated to provide a single point of contact for prompt resolution of service level defaults and any other failures regardless of whether the service level failure was caused by the supplier, the customer, or by a third party.

Service Level Provisions Commonly Found in a Service Level Agreement or Attachment

While the terms described above are commonly located in the main body or terms and conditions of an IT agreement, the specific details with respect to the service levels are more commonly located in a separate service level agreement or attachment that is commonly attached to the main agreement as an exhibit. The terms below are commonly found in a service level agreement.

Measurement Window and Reporting Requirements

In each service level agreement, it is important to specify how frequently service levels are to be measured and reported. As a general matter, consider specifying that, for example, service levels are to be measured on a monthly basis and a report is to be delivered to the customer by the supplier with a few days (no more than ten) after the end of each month. This is commonly referred to as the service level measurement window. With respect to reporting, be as specific as possible. Service level agreements commonly describe how many copies of each report are to be delivered and the format (e.g., Microsoft Excel) of the report. Customers typically retain control over the reporting requirements and should be permitted the opportunity to change the requirements upon notice to the supplier.

- Prior to any requirement that the supplier meet the service levels, the supplier should be required to implement and have in place the appropriate measuring tools to equip the supplier with the ability to measure its performance against the service levels. Failure to implement such measuring tools will impact the supplier's ability to report on whether the services are actually being provided as required.

- The service level agreement should clearly state the reporting requirements such as:
  - hard copy, soft copy, and online reporting requirements;
  - when each report must be delivered;
  - to whom each report must be delivered;
  - whether multiple copies of reports are required;
  - whether supporting information is required to substantiate each report; and
  - whether summary reports are required and the frequency of such reports (e.g., quarterly or annually)

Maximum Monthly At-Risk Amount

Service level agreements frequently contain terms that cap the supplier's liability for service level or performance credits at a percentage of the fees paid by the customer to the supplier. For example, the parties may agree that the performance credits in any given calendar month will not exceed the fees paid or to be paid by the customer to the supplier in that month. In larger IT transactions, the parties may agree that the total at-risk amount for each month is a percentage (e.g., 25%) of the fees paid or to be paid by the customer to the supplier in the month in which the services failed to meet the service levels.

Performance Credits

Performance credits are a credit to which the customer in an IT transaction is entitled to in the event that the supplier fails to perform the services in accordance with the service level to which the performance credit is assigned. Performance credits generally reflect, in part, the diminished value for the services delivered as compared to the service levels and other contractual commitments and generally do not represent damages, penalties, or other compensation remedy that may result from a supplier's failure to meet the service levels and other contractual commitments. Care should be taken to avoid service levels as a customer's sole and exclusive remedy for the supplier's failure to provide the services in accordance with the service levels and other contractual commitments. Often, performance credits are a specified amount (e.g., $500) or a specified percentage (e.g., 25%) of the fees paid or to be paid for the particular service that is the subject matter of the service level failure. For example, in a simple hosting transaction, you might see three to five service levels focused on:

- download time during critical hours;
- download time during noncritical hours;

- availability during critical hours;
- availability during noncritical hours;
- response time; and
- incident resolution time.

For each of these categories, a service level and a credit would be assigned. The credit could be a specific dollar amount the customer would be entitled to in the event the supplier failed to meet the service level, or it could be a percentage of fees to be paid to the supplier during the measurement window (e.g., a month or a quarter).

In larger and more complex IT transactions, it is appropriate to develop total category allocation pools. The parties would develop a total pool (e.g., 200%) and then assign a portion of the pool to the service levels. This portion of the allocation pool assigned to each service level operates as a weighting factor. In these cases, it is appropriate to establish upward and downward limits on the portion of the allocation pool that can be assigned to each service level. For example, in some cases, it would be appropriate to limit the assigned amount to no more than 25% and not less than 5%. In this way, the parties can allocate the risk associated with service level failure appropriately and fairly. Ultimately, when this more complex approach to performance credits is used, the actual performance credit will be determined mathematically by multiplying the assigned performance credit allocation percentage by the total monthly at-risk amount. The service level agreement should clearly describe this methodology and, where appropriate, use examples. For example, the following could be included:

> For each Critical Service Level Failure, Supplier shall pay Customer a Performance Credit that will be computed in accordance with the following formula:

$$\text{Performance Credit} = A \times B$$

where

A = Performance Credit Allocation Percentage
B = Monthly At-risk Amount

For example, in a month that supplier fails to meet a critical service level, assume that (i) supplier's monthly charges (i.e., base monthly charges for all services) during the month in which the critical service level failure occurred were $500,000, (ii) the maximum monthly at-risk percentage is 14%, and (iii) the performance credit allocation percentage for such critical service level is 20%. The performance credit due to customer for such critical service level failure would be computed as follows:

A = 20% (the Performance Credit Allocation Percentage) multiplied by
B = $70,000 (the Monthly At-Risk Amount = 14% of $500,000)
= $14,000 (the Amount of the Performance Credit)

In addition to the foregoing, it is essential the service level agreement specify in detail all of the requirements with respect to performance credits. It is frequently the case that service level parameters are heavily negotiated between the parties to the transaction. Topics to consider include the following:

- What will be the limits on performance credits, if any, if more than one service level failure occurs in the same measurement period? The best-case scenarios for customers is that the sum of all performance credits applicable would be credited to the customer, even if multiple service level failures occur within the same measurement window. Suppliers will want to limit recovery and may require that only the highest of the performance credits apply. In cases where there is a monthly at-risk amount, performance credit recovery would be limited to the monthly cap.
- Notification of service level failure frequently comes from the supplier, since the supplier will often be the first, and sometimes only, party that becomes aware of the failure. Timing requirements for notification are essential. A specific timeframe (e.g., no more than fifteen minutes after becoming aware of the service level failure) is preferable, but if the supplier won't commit to that, a broader timeframe (e.g., promptly after becoming aware of the service level failure) can be used.
- The parties will want to determine ahead of time whether performance credits are to be the customer's sole and exclusive remedy associated with the supplier's failure to provide the services in accordance with the service levels.

Presumptive Service Levels

It is frequently the case when negotiating an IT transaction that insufficient data exists to determine the appropriate standard for service. In such a case, it is common for the parties to agree to a measurement study period that is used to establish the appropriate standard. In such a case, the following terms and responsibilities should be included in the contract:

- The supplier should be required to begin measuring its performance against the presumptive service levels on an agreed-upon date following initiation of the supplier's provision of the services.
- During the measurement period, the parties must monitor and analyze the supplier's performance against the presumptive service levels.

- During the measurement period, the parties should meet to discuss the presumptive service levels given all relevant factors such as the customer's business needs, the supplier's performance during the measurement period, applicable industry standards with respect to the presumptive service level, nonrecurring events responsible for degradation in service, and steps that both parties have taken and could take to improve service level performance.
- At the end of the measurement period, the presumptive service level will be changed to a set service level determined by the parties.
- The contract should contain a formula and process for converting presumptive service levels to set service levels. For example, the set service level could be the midpoint between the presumptive service level and the supplier's actual performance during the measurement period. If the presumptive service level is 96% and the supplier achieved a 98% performance during the measurement window, the service level would, in this example, be 97%. In many instances, this is not appropriate, for example, where the supplier's performance never dipped below 98% during the measurement window. Then, the customer has a good argument that the set service level should be no less than 98%. As a result, the contract should include the opportunity for each party to challenge the result of the conversion formula if a party can establish that the formula does not establish a reasonable result given all relevant factors.

Exceptions to Service Levels

In most cases, suppliers will want to ensure that certain failures to meet the service levels are excused. These cases usually focus on the following types of causes of service level failure:

- The customer's provided equipment, telecommunications, or software failures that are beyond the control of the supplier
- The customer's failure to implement changes in systems, hardware, or software that are reasonably required and for which the customer has been provided a significant amount of notice
- A significant third-party event that causes service delivery failure
- A force majeure event
- A disaster in which disaster recovery and business continuity obligations are not a required element of the transaction

Care should be taken to review all exceptions to performance carefully to ensure they are properly tailored and not overly broad.

While almost every service level agreement contains an exclusion for force majeure events, this does not mean that a customer should be left with no remedy. When such an exclusion is present, the customer should modify the force majeure provision along the following lines: "In the event a force majeure event impacting Vendor's provision of the Services continues for a period of more than five (5) business days, Customer may terminate this Agreement by providing written notice to Vendor. For the avoidance of doubt, in the event Vendor's provision of the Services is the subject of a force majeure event, the fees to be paid by Customer hereunder shall be equitably adjusted to reflect the period in which performance was impacted."

Supplier Responsibilities with Respect to Service Levels

It is important to understand that suppliers have control over their performance and will typically be the first to know if there is a service deficiency or failure. It is imperative that service level obligations include the supplier's responsibility to promptly notify the customer when problems arise and quickly initiating investigations, root cause analyses, and remediation efforts. What's more, if the supplier becomes aware of a potential problem or anticipated interruption in the provision of service or the supplier's ability to provide the services in accordance with the services levels and the agreed-upon standards, the supplier should be obligated to notify the customer and take all efforts necessary to avoid the interruption or degradation in service. Written recommendations for improvement in the services to avoid future degradation in the supplier's performance are also frequently required.

Additions, Deletions, and Modifications to Service Levels

Most service level agreements will include a process for adding, deleting, and modifying service levels over the course of the relationship.

- Additional service levels can typically be added to a service level agreement after a period of time has passed (e.g., six or twelve months). Where the parties are unable to agree to the level of performance required for the new service level, a methodology will be included in the service level agreement. For example, the parties could agree that the service level will be defined as the average of service measurements for the most recent three consecutive calendar months that the supplier has been providing the particular service. Where historical data does not exist, the parties could agree to measure the service for a specified period of time and then, at a later date, determine the appropriate service level by working together to assign the service level or, where appropriate, by taking an average of the data collected over the measurement period. In cases where an allocation pool is used to determine performance credits, as discussed previously, the pool would require redistribution in order to accommodate the new service levels.

- Customers typically have the right to delete certain service levels, also after an agreed-upon period of time (e.g., six months after the effective date). In cases in which an allocation pool is used to determine performance credits, as discussed previously, the pool would require redistribution in order to accommodate the deletion. The customer would typically be able to redistribute the allocation in its discretion and following any guidelines agreed to by the parties (e.g., maximums and minimum allocation parameters).
- In many cases, it is also appropriate to include a procedure for the customer to modify the allocation pool across the service levels. Where a two-tiered service level structure is used (i.e., using key performance indicators that typically wouldn't carry performance credits and critical service levels that typically would carry performance credits), this procedure would also include a process for the customer to "promote" key performance indicators to critical service levels. The customer's ability to modify the distribution of the allocation pool and to promote key performance indicators to critical service levels is commonly limited to during the term (e.g., no more than twice annually during the term). In the event of modifications to the allocation pool, any agreed-upon minimums and maximums would continue to apply. In the event of a promotion of a key performance indicator to a critical service level, the customer will need to redistribute the total allocation pool across critical service levels to ensure that the total does not exceed the total performance credit allocation percentage.

Earn-Back

In some transactions, it is appropriate for the supplier to have the opportunity to be repaid, or "earn back," performance credits that were paid out to the customer for the supplier's failure to provide the services in accordance with the service levels. If an earn-back opportunity is appropriate, it is essential that the service level agreement include a procedure with respect to how the supplier can achieve the earn-back.

- Consider whether the earn-back opportunity should apply to all of the service levels or if certain service levels (like the service levels that measure the most critical services that the supplier is going to be providing or service levels that measure the services being provided at critical times like, in a hosting transaction, availability of applications during peak periods) should be excluded from the supplier's opportunity to earn back performance credits.
- At the end of a measurement period (e.g., at the end of each year), the supplier must be required to report on at least the following information:
  - The supplier's actual average monthly performance during the measurement period.
  - The number of months in the measurement period that the supplier met or exceeded the required service level.
  - The yearly performance average for each service level.
  - The total amount of performance credits accrued and credits for service level failures.
- If the yearly average for a service level meets or exceeds the required service level, then the supplier may be relieved from paying any performance credits that would otherwise be paid for the particular service level.
- If the yearly average for a service level falls below the required service level, the supplier is required to pay all performance credits due to the customer within a specified period of time. Payment of performance credits could come in the form of a credit on the next invoice sent by the supplier to the customer, or if there are no further invoices to be issued, a monetary payment should be required.

Form of Service Levels

Included below is a commonly used approach to mapping service levels in a service level agreement. While many forms of service levels are appropriate, it is essential that each service level contain all of the necessary information with respect to the service level such as the title of the service level, the measurement window, the actual service level (or in the example below, the expected service level and the minimum service level), the calculation for how the service level is derived, and any additional information or description that will aid the parties in meeting, exceeding, and calculating the service level.

No.	*Title*	*Performance Credit Allocation Percentage*	*Description*	*Service Level*	
1. Mainframe computing operations					
C1.1	Mainframe production environment availability	10%	The availability of the mainframe production environment as described in Exhibit E.4.a (mainframe applications portfolio), in each calendar month	Measurement window Expected service level Minimum service level Calculation	Calendar month 99.98% or greater availability or greater in each calendar month 99.96% or greater availability or greater in each calendar month Availability of mainframe production environment in each calendar month
2. Distributed computing operations					
C2.1	System high availability	10%	The availability of each OS instance—high availability as described in Exhibit E.4.d (server architecture standards) in each calendar month	Measurement window expected Service level Minimum service level Calculation	Calendar month 99.97% or greater availability in each calendar month 99.95% or greater availability in each calendar month. The availability of OS instances—high availability in each calendar month

Summary

If used properly, service levels can help to ensure that a supplier's delivery of services meets and exceeds customer's expectations. The pricing structure of many IT transactions assumes a specified level of performance. If the supplier's performance of the services fails to meet the customer's expectations, as articulated in a service level agreement, a performance credit structure can help to ensure that the customer is only paying for the value of the services actually received. To that end, service levels can help incite and maintain supplier performance. To work appropriately and fairly to both parties, a detailed process and procedure should be implemented to ensure the appropriate application of service levels to a customer's commercial transactions.

Chapter 10

Idea Submission Agreements

CHECKLIST

Understanding Risk of Submissions

- ☐ Determine whether you collect "submissions"
- ☐ Implement an idea submission agreement
- ☐ Determine if purchase or license agreement is required
- ☐ No compensation for submissions (unless provided in an agreement)
- ☐ No confidentiality of submissions (except under an NDA or as required by an idea submission agreement)
- ☐ No submission of ideas without idea submission agreement (e.g., via e-mail)

Key Provisions of Idea Submission Agreement

- ☐ Simple agreement or "full-blown" agreement
- ☐ No compensation
- ☐ No confidential treatment
- ☐ Writing requirement
- ☐ Demonstration of IP rights
- ☐ No obligation to return

- ☐ No obligation to provide any confidential information to the other party
- ☐ Ability to contest IP rights in submissions

Reverse Submissions

- ☐ Avoid broad feedback provisions

Overview

Every business has some form of product or service that it offers to its customers and business partners. In general, those products and services are comprised of various types of intellectual property. As is frequently the case, the business's customers, business partners, and even third-party vendors may make suggestions, provide ideas, or offer feedback that may be useful to the business in improving its products and services. This type of information is sometimes generally referred to as a "submission." While submissions can be an excellent means toward product and service improvement, the unsolicited nature of this information can also create intellectual property ownership issues. That is, the submission may itself constitute intellectual property.

Businesses must be careful to ensure their product development efforts are not contaminated by outside intellectual property. If the business allows its developers to be exposed to submissions from the outside, those outsiders may later claim the business did not have authorization to use their intellectual property. They may demand license fees and other compensation for use of their submissions.

This type of risk occurs all the time in the movie and television industries where people constantly offer ideas for new shows to the studios. If the studio later creates a show similar to one of those ideas, even though it developed the show completely without reference to the idea, they could still potentially be sued for misappropriation of the idea. The studio would then be in the position of needing to prove a negative: that it did not use the idea in the creation of its program.

Businesses that receive outside submissions run a similar risk. To minimize the potential for disputes and litigation, businesses must implement measures to control how submissions are made and ensure they have proper rights to use the submissions if they choose to do so.

Key Risks of Submissions

The first step in developing an approach to managing submissions is to define what they are and appreciate their inherent risks.

- Often businesses will solicit or receive unsolicited ideas, feedback, and submissions ("submissions") from customers, suppliers, business partners, and other third parties. Submissions can provide valuable and important information for the business, but they can present significant risks—particularly with regard to intellectual property ownership.
- The threat to intellectual property can be significant. If a submission is already under development by the business or would have been arrived at in the ordinary course of the evolution of the business' products and services, its development process may be irretrievably tainted. That is, the business will be placed in the position of having to prove it independently arrived at the idea without reference to the submission. This can be a difficult hurdle to overcome and could require litigation to resolve.
- If the business obtains ideas or other intellectual property from third parties and subsequently modifies or improves that intellectual property, the modifications or improvements may be owned by the third party or jointly owned by the business and the third party. These situations should be avoided if at all possible and can be managed by using an appropriate "idea submission agreement."
- Some third parties may expect compensation for their submissions. If the company accepts submissions without a clear understanding regarding compensation, it may face a claim from the third party for compensation it did not intend on paying. In the event the business desires to purchase or acquire patentable ideas, copyrighted materials, or other intellectual property from the third party, this must be accomplished through use of a separate agreement, usually a purchase or license agreement.
- Accordingly, the business must not accept submissions without a clear understanding about whether the third party will be entitled to compensation. Typically, businesses only accept submissions on the condition that the third party agrees it is not entitled to any compensation, except in a narrow range of instances in which the submission is provided under an express understanding that compensation will be paid.
- Third parties may want their submissions to be treated confidentially, or assume that the business will keep their ideas and materials confidential. If submissions are accepted without a clear understanding regarding confidentiality, the business may find itself unknowingly bound by implied confidentiality obligations.
- Accordingly, the business should not accept submissions subject to any confidentiality restrictions unless it has made an informed decision that the submitter has valuable confidential information that the business is comfortable receiving subject to confidentiality restrictions. If the business is willing to accept submissions under an obligation of confidentiality, an appropriate nondisclosure agreement should be negotiated. In all other situations, the

idea submission agreement should make clear that submissions are not subject to confidentiality restrictions.

- Common sources of submissions are e-mails and communications made through business websites. Submissions in the form of e-mails should be returned to the sender with a clear statement that submissions cannot be considered without an appropriate idea submission agreement. Submissions presented through websites should be addressed by the addition of language to the relevant website terms and conditions, making clear the conditions under which submissions may be made.

Essential Terms

We have already discussed several key elements to be included in any idea submission agreement (e.g., compensation and confidentiality). Several additional areas should also be addressed in agreements governing submissions. Note that a "full-blown" idea submission agreement is likely not warranted in many situations. In those situations, a brief paragraph may be included about submissions as part of an overarching contract or in website terms and conditions.

- Submissions should be made in writing. If the submission is not in writing, it may not be clear what was submitted, resulting in uncertainty and risk in the event of a claim for compensation by the third party. Verbal submissions should be avoided to the maximum extent possible.
- In most cases, the third party should be required to disclose any patents, copyright registrations, or other documentation evidencing its intellectual property rights relating to the submission.
- The business should not have any obligations to return any submissions or related documentation.
- The idea submission agreement should clearly state that the third party will be entitled to no compensation for the submission.
- The idea submission agreement should clearly state that the submission will not be treated as confidential information. If the situation warrants confidential treatment, then an appropriate nondisclosure agreement should be used.
- The third party should represent and warrant that it has the right to provide the submission at no cost and under no obligation of confidentiality.
- The idea submission agreement should make clear that the business has the right to use or disclose the information as it deems appropriate in connection with its business.
- The business should be under no obligation to provide any of its own confidential information, trade secrets, research, test results, or other information to the third party.

- The idea submission agreement should clearly state that the business reserves all rights to contest the validity and enforceability of any patents, copyrights, or other intellectual property rights claimed by the third party. For example, although the third party may have a patent, it may not have been validly issued.

Beware of Reverse Submissions

Businesses must also be cautious of situations in which they are asked to accept another company's submission terms (i.e., the reverse situation of the one discussed in the previous section). This occurs most often in the context of software licenses and other technology agreements. In those cases, the business is licensing a third party's intellectual property. As part of that license, the contract may include some form of "feedback" or submission language, granting the third party broad rights in any suggestions or feedback the business may offer relating to the third party's products. These provisions should be carefully scrutinized and narrowed to ensure the business does not give up any of its preexisting intellectual property rights and to limit the provision to only information specifically directed at the third party's product. In some instances, it may be appropriate to request the language be made mutual such that each party can benefit from submissions regarding its respective products and services (e.g., in the context of a lengthy software implementation engagement, the licensor may make suggestions to improve the licensee's operations).

Summary

Submissions can be an excellent way to gain valuable information and feedback from customers, business partners, and others about your products and services. Caution must be taken, however, to ensure that information does not taint your development process or expose your business to infringement claims or demands for compensation. Finally, care should be taken whenever you are asked to accept submission language in a contract offered by one of your vendors. Train your personnel to be on the look for situations where submissions may be made and ensure the procedures described in this chapter are implemented.

Chapter 11

Joint Marketing Agreements

CHECKLIST

Preengagement Considerations

- ☐ Scope of engagement
- ☐ Nondisclosure agreement

Marketing Obligations

- ☐ Exhibit with precise marketing obligations
- ☐ Mutual agreement
- ☐ Responsibility for expenses—shared

Referral Arrangements

- ☐ Definition of who is a referral
- ☐ Referral period (anything outside of period is not a referral)
- ☐ Compensation to referring party
- ☐ Audit rights

Confidentiality

- ☐ Protection of all confidential information exchanged
- ☐ Overrides any NDA entered into in preengagement

Intellectual Property Issues

- ☐ Any software or products are provided as is
- ☐ Return of software/products at end of relationship
- ☐ Narrow license for any materials or information shared
- ☐ License for use of trademarks and names
- ☐ Approval for trademark/name usage
- ☐ Reservation of IP rights
- ☐ Residual knowledge/feedback clause

Warranties and Disclaimers

- ☐ Basic warranties
  - Ability to enter into agreement
  - Compliance with applicable laws
  - No pending or threatened litigation
- ☐ Broad warranty disclaimer
- ☐ No guarantee of revenue (unless appropriate)

No Agency

- ☐ Agent relationship not intended

Limitations of Liability

- ☐ Common limitation of liability exclusions
  - Breaches of confidentiality
  - Claims for which the vendor is insured
  - Indemnification obligations
  - Infringement of IP rights
- ☐ Disclaimer of all other liability

Indemnification

- ☐ Limited to violations of law and misuse of IP

Term and Termination

- ☐ Renewal after stated term
- ☐ Free ability to terminate
- ☐ Termination for breach

Overview

Joint marketing agreements can take several forms. There is no one-size-fits-all for this type of contract. Generally speaking, joint marketing agreements are used in transactions in which two or more parties come together to jointly market a product or service. In some cases, the agreement can be limited to a simple contract for joint marketing and promotional efforts whereby the parties to the contract jointly market a product or service and agree to a revenue share arrangement for fees collected as a result of such efforts. Frequently these relationships are expanded to include referral arrangements in which one or all parties also refer potential customers to the other party or parties in exchange for monetary or other compensation. These types of agreements may also include product integration obligations and terms for jointly supporting mutual customers. Depending on the extent of those additional terms, separate, more detailed agreements may be required to fully address the parties' respective integration and support obligations. However, in many instances, these agreements can be very simple and entered into to "get a deal off the ground." They provide the basics and provide the framework for the parties to grow the relationship in the future.

Since joint marketing agreements generally reflect many, if not all, obligations as mutual, they force both parties to be reasonable in the language they may request. Knowing that a requested provision will be made mutual will cause the drafting party to be reasonable from the outset. This inherent "mutual assured destruction" aspect of joint marketing agreements can make them very easy and quick to negotiate.

Key Considerations and Essential Terms

Since the parties to a joint marketing agreement will be working closely together, two primary concerns arise. First, the parties will be sharing information about their products and, therefore, each will be concerned about protection of its intellectual property. Second, the parties will want to avoid any inference that is seen as either the other's partner or agent, designations that can raise significant legal issues.

When drafting and negotiating a joint marketing agreement, consider these terms to ensure that the drafting and negotiation process is efficient and that your company's business and legal objectives are achieved.

Determine the Scope of the Engagement

- Because joint marketing relationships can take one of many different forms, the first step will always be to determine the scope of the engagement and ensure that the agreement covers all relevant activities in which the parties will engage.
- Depending on the nature of the information to be shared between the parties, a nondisclosure agreement should be entered into prior to commencing

substantive discussions. Critical confidential information may exchange hands prior to drafting and negotiating a definitive joint marketing agreement. You will want to ensure that the appropriate protections are in place with respect to that information. As discussed in Chapter 3 (Nondisclosure Agreements), the confidentiality provisions of the joint marketing agreement will take precedence over any confidentiality agreement entered into during negotiations.
- A common error in these engagements is attempting to do too much in a single agreement. Except in instances where a highly specialized contract is used, joint marketing agreements should not be used for joint development work, significant engagements where the parties will integrate their support organizations, or instances in which the other party is in reality serving as a reseller. In those cases, a development or reseller agreement is more appropriate.

Marketing Obligations

- The greatest source of disputes and potential litigation arising from joint marketing agreements are misunderstandings about the parties' respective obligations to market and promote the relationship. These types of disputes arise more frequently where the agreement contains only vague or ambiguous marketing obligations (e.g., the parties will use their best or commercially reasonable efforts to promote the good or service being marketed). These types of provisions should be rejected in favor of specific language with respect to the obligations of each party to promote the good or service that is being marketed (e.g., each party shall include promotional materials for the good/service in all of the products that it sells to customers) or as the parties "may mutually agree upon from time to time." In addition, it is always a good idea to agree on specific marketing efforts in which the parties will engage and to include those efforts in an exhibit to the agreement.
- A common mistake is to attempt to include too much detail about the parties' intended marketing obligations in the body of the agreement. The specifics about those efforts should be set forth in an exhibit. The exhibit can describe everything from joint press releases, to joint participation in seminars and industry events, to the development of hyperlinks between the parties' websites, to approved content for advertising, to the number of advertisements that will be conducted each month.
- If the parties are unable to specify an intended marketing activity in detail, it should be written as "subject to the parties' mutual agreement." This will minimize the potential for disputes over an obligation that cannot be readily described in detail at the time of contract execution.
- Unless the parties have agreed otherwise, express language should be added to the agreement making clear each party will bear their own expenses in engaging in the marketing activities.

Referral Arrangements

- Many joint marketing agreements include language governing referrals made from one party to the other of prospective customers. These provisions may be for the benefit of either one party or both.
- Referral provisions should clearly define who is a "referral" and who is not. For example, referrals should clearly be defined to exclude any individuals, entities, or organizations who are current customers of a party or who have been independently contacted by the party independent of the referral or joint marketing relationship.
- The referral provision should always specify a period (usually twelve months) following the date of the referral after which no compensation will be due to the referring party if the referral later purchases products. For example, most agreements impose a time limit of twelve months following the date of the referral. If the prospect becomes an actual customer during that time and purchases products, the referring party will be due their compensation. However, if the prospect does not become an actual customer until after the twelve-month period has lapsed, the referring party will be due no compensation.
- In general, compensation for the referral is paid to the referring party. That compensation is usually based on some percentage of the net revenue derived from the sale of the products to the referral. Other compensation models, such as a fixed fee, are also possible. In any event, compensation is normally based only on the sale or license of products, not professional or support services.
- Audit rights should be included in every agreement in which referral compensation will be paid. This is to ensure the amount of referral compensation is accurately computed. If only one party will be the referring party, a standard audit provision should be included that provides for shifting of the cost of the audit to the other party if the audit identifies an underpayment of more than a specified amount (e.g., 5%–15% of the fees due). If the referral arrangement involves referrals of prospects from the other party to your company, it is in your best interests to limit onerous audit provisions, particularly those in which the other party or its auditors are granted the right to access your facilities and systems. In those cases, it is often better to include a less intrusive obligation to simply provide reasonable documentation regarding the calculation of the referral compensation.

Confidentiality

Since the parties will likely be exchanging product information, marketing plans, customer names, and other sensitive information, the agreement should include a strong confidentiality clause. As discussed in Chapter 3 (Nondisclosure Agreements), the provision should make it clear that it takes precedence over any prior nondisclosure agreements entered into by the parties and that such prior agreements are terminated in lieu of the confidentiality provisions of the joint marketing agreement.

Intellectual Property Issues

- If any software or other products are exchanged for purposes of training, demonstrations, and other marketing activities, the software and products should be provided under their standard commercial licenses and product terms and conditions. Since the software and products are not being provided as the result of a normal license or purchase transaction, the marketing agreement should make clear the software and products are provided on an "as-is" basis. The software and products should be returned on expiration or termination of the marketing agreement or at any time on the providing party's request.
- In addition to providing product samples, the parties to the joint marketing agreement may provide other materials, information, and content for use in the joint marketing activities. All of these items likely constitute intellectual property and confidential information of the providing party and should, therefore, be provided under a narrow license permitting their use only in connection with the joint marketing activities, as authorized by the providing party.
- Since most joint marketing activities will involve the use of one or both of the parties' names and trademarks, every joint marketing agreement must include a specific license governing use of those names and trademarks. The license is generally written as a "cross-license," meaning that each party grants the other party a license to its names and trademarks.
  - Avoid trademark licenses that include all of a party's trademarks. Instead, a license should be granted only to those trademarks a party specifically identifies as being subject to the license and necessary to support the joint marketing efforts of the parties.
  - Trademark licenses must include provisions requiring the approval of the trademark owner before any use is made of its trademarks.
- The joint marketing agreement should also contain a clear statement that each party reserves all rights in its intellectual property and that neither party is transferring or assigning any rights in its intellectual property. This will assure each party that it has the appropriate protection of its intellectual property during the relationship and after any termination or expiration of the agreement. They key concept is that neither party should be placing their intellectual property at risk by virtue of an engagement of this kind. This does not, however, mean that the parties might agree to enter into a professional service agreement at a later time that would provide for the transfer or assignment of intellectual property rights.

- Since the parties will be working closely together and discussing their mutual products, there is the potential for corruption of a party's development process. If the other party suggests an idea already under development or that would have been arrived at in the ordinary course of the evolution of your products and services, your development process may be irretrievably tainted. That is, your company will be placed in the position of having to prove it independently arrived at the idea without reference to any information provided by the other party. This can be a difficult hurdle to overcome and would require litigation to resolve.
- As a baseline means of mitigating this threat, it is common to include a residual knowledge and feedback clause. This clause permits a party to continue to use the information its employees retain in their unaided memories, provided the information does not infringe the other party's intellectual property rights. It also addresses the situation in which one party specifically makes suggestions to the other party for the improvement of its products.

Warranties and Disclaimers

- The joint marketing agreement should include basic warranties ensuring that the parties have the ability to enter into the agreement and will comply with applicable law. Other warranties may be appropriate in particular engagements (e.g., noninfringement warranties, antivirus warranties).
- Apart from the foregoing basic warranties, joint marketing agreements commonly include strong warranty disclaimers and limitations of liability.
- Unless the agreement includes specific revenue commitments, a disclaimer should be added making it clear that neither party is guaranteeing the other party will derive any specific revenue from the relationship. While revenue commitments may be acceptable in some relationships, if they are not applicable, it is important to make that clear in your joint marketing agreement.
- **No agency.** Since the parties will be working closely together and participating in joint efforts, there is the possibility the relationship may be construed as a partnership or engagement in which each party is acting as the other party's agent. If this is not the case (and it is usually not), the joint marketing agreement should state clearly that such an agency relationship is not intended.
- **Limitations of liability.** Apart from standard carve outs to any limitation of liability (e.g., damages arising as a result of a party's breach of its confidentiality obligations, damages arising out of a party's misappropriation of the other party's intellectual property, damages arising out of or payments to be made pursuant to a party's obligation to indemnify the other party), the agreement should disclaim essentially all other liability for both direct and consequential damages. If referral fees will be paid, it is common to limit liability to a

multiple of fees paid over a specified period of time (e.g., the amount of fees paid in the three months prior to the incident that gave rise to the claim).

- **Indemnification.** A mutual indemnity is normally included to protect against violations of law or misuse of the other party's intellectual property. As noted above, it is common that damages arising out of and payments to be made pursuant to a party's obligation to indemnify the other party are carved out of (i.e., excluded from) any disclaimer of consequential damages, cap on direct damages, or any other limitation of liability.

Term and Termination

- The agreement may contain a term, which may be a particular period of time (e.g., one year). Following the initial term, the agreement would typically automatically renew for additional one-year terms unless either party gives notice of its intent not to renew.
- Unless business reasons dictate otherwise, the parties typically retain the right to terminate a joint marketing agreement for its convenience (i.e., at any time and for any reason or no reason at all) on written notice to the other party.
- In most cases, either party should have the right to terminate a joint marketing agreement if the other party breaches the agreement.

Summary

Joint marketing agreements are used in cases where two or more parties come together to jointly market and promote a product of one of the parties or a product or service that the parties have jointly developed. In order for joint marketing relationships to be successful, it is essential that each party shares information about its products with the other party. As a result, the parties' primary concern will be to protect their intellectual property rights. Where referral arrangements are agreed to in joint marketing relationships, the parties will want to be specific about the scope of the referral requirements and the revenue share that is required. Care should be taken to ensure that if a joint marketing agreement includes development-like or resale-like requirements, then the appropriate relationship be structured.

Chapter 12

Software Development Kit (SDK) Agreements

CHECKLIST

Content of SDK

- ☐ APIs
- ☐ Sample code
- ☐ Sample documentation
- ☐ Other data and information
- ☐ Ensure IP protected

Scope of License

- ☐ Scope
- ☐ Internal development
- ☐ Internal testing
- ☐ Distributable
- ☐ What agreement applies to distribution
- ☐ Developer rights

Ownership

- ☐ Company owns SDK materials
- ☐ Modifications and derivatives
- ☐ Company ownership of suggestions and feedback

Confidentiality

- ☐ Testing of third-party software products
- ☐ No representations or warranties as to compatibility
- ☐ Confidentiality of SDK

Support

- ☐ Not generally provided
- ☐ If provided, precise obligations
- ☐ No representations or warranties with respect to support services
- ☐ As is and as available

Warranty Disclaimers

- ☐ No warranties
- ☐ No liability of company
- ☐ As is, as available

Limitations of Liability

- ☐ Complete limitation
- ☐ Exclusion of consequential damages
- ☐ No recovery of direct damages
- ☐ Stop gap if unenforceable

Indemnification

- ☐ Developer indemnification of company for all claims against company related to developer's use of SDK

Export and Import

- ☐ Developer's compliance with all export and import laws
- ☐ Indemnification for claims against company
- ☐ Acquisition by federal government
- ☐ Protection of IP

Term and Termination

- ☐ Term?
- ☐ Termination for convenience
- ☐ Termination for breach/bankruptcy

Overview

A software development kit (SDK) is comprised of software tools and documentation that enable a third-party software developer to interface with a company's software and/or hardware products; but does not contemplate the redistribution of any company software or products to third parties (i.e., a reseller or original equipment manufacturer (OEM) agreement would be required for any redistribution). SDKs must be provided pursuant to an SDK agreement that provides critical protections for the company's intellectual property and ensures that the company is not exposed to undue risk as a result of permitting access to the company's software and/or hardware products.

SDKs are generally furnished as-is under a nonnegotiable simple agreement. The party desiring to use the SDK must either accept the agreement without modification or, potentially, forego the opportunity.

Key Considerations and Essential Terms

SDK agreements must generally contain the following terms in order to ensure that the company's business and legal objectives are achieved.

- The SDK should include enough data and information, such as application programming interfaces ("APIs"), sample code, and documentation, to accomplish the goal of allowing third parties to develop applications to interface with company software and hardware. However, company intellectual property is a critical concern, and the company should not provide more than is necessary to accomplish this goal, particularly with respect to the company's source code and other trade secrets.

Scope of License

- SDK agreements should clearly specify the scope of the license being granted and what the third party can do with the SDK materials, such as APIs, sample code, and documentation. For example, certain code may be for internal development and testing only, while other code may be distributable in object code format only, but only under a separately executed OEM or reseller agreement.
- The agreement should describe what products or services will be provided by the developer for interfacing with the company's software and/or hardware.

Ownership

- The company must retain ownership of all right, title, and interest in the SDK materials, and all intellectual property rights therein, including any modifications or derivatives of the SDK materials. Each SDK agreement should clearly establish the rights of ownership to the SDK materials. In addition, the company should retain the right to use and disclose any suggestions or feedback provided by the developer relating to the SDK materials or any of the company's software or hardware products. That is, if the developer makes a suggestion for the improvement of the SDK materials, the company should have unlimited rights to use and implement that suggestion in its business.

- In many instances, the company should consider including protection to ensure that if a third party uses the SDK to create something that the third party will not later sue the company if another licensee of the SDK chooses to do something similar or, if the company, itself, decides to do something similar. This language generally takes the form of a "non-assert," meaning that the licensee will not attempt to assert intellectual property rights it develops using the SDK against the original company or its customers.

Confidentiality

- Each SDK agreement should contain strong confidentiality restrictions and protections with respect to the SDK materials, particularly if any source code will be disclosed to the other party.

Compatibility Testing

- Depending on the circumstances, the company may want to require the right to test the third party's software/products to confirm they are compatible with the company software and/or hardware. In these circumstances, the language should clearly state that the testing is only for the company's purposes and benefit and is not a representation or warranty to the developer or any third party (including end users) with respect to the developer's products and/or compatibility with company's software and hardware. In certain instances, the company may want to develop a "seal" or "certification" program under which third-party products may expressly be certified by the company as compliant with its products. These programs, however, must be carefully constructed to ensure that the company does not assume liability to end users who purchase the third party's products and later encounter a compatibility issue.

Support

- Support is not typically provided with the SDK. If the company is to provide support, the support obligations should be precisely defined in the SDK agreement. Any services of this kind are generally provided on a completely as-is and as-available basis.

Warranty Disclaimers

- Since SDKs are typically provided to developers at no charge, the SDK agreement should contain very strong warranty disclaimers and limitations of liability. The SDK materials should be provided completely "as-is" with no warranties, express or implied.

Limitations on Liability

- Companies are not generally liable for any damages whatsoever with respect to SDKs, and the SDK agreement should make this clear. Because SDKs are provided at no charge, it would be unusual and inappropriate for the company to assume liability for the SDK. This limitation of liability should apply to both consequential damages (e.g., lost profits, data loss) and any other damages, including any direct damages suffered by the developer.
- It is important to provide a stop-gap/fail-safe measure in the event any portion of the limitation is held unenforceable (e.g., the agreement should provide a very low liability cap, such as $50).

Indemnification

- The developer should agree to indemnify and hold the company harmless from any and all losses and third-party claims against the company relating to the developer's use of the SDK, or any software, hardware, or services provided by the developer. The indemnity should include protections from claims by any third party to whom the developer may distribute its products and services.

Export/Import

- In light of certain restrictions on the export and import of software and other technology, SDK agreements should require the developer to comply with all export and import laws and to indemnify and hold the company harmless from any claims arising out of the developer's breach of such laws.

Acquisition by Federal Government

- Because the federal government can obtain certain rights to software provided to the government, the SDK agreement should include language protecting the proprietary nature of the company's software.

Term and Termination

- While not always necessary, SDK agreements may contain a term, which may be a particular period of time (e.g., one year). Unless business reasons dictate otherwise, the company should have the right to terminate the SDK agreement for its convenience (i.e., at any time, for any reason, or for no reason at all, immediately and without cause). Written notice of termination (e.g., ten business days prior to the effective date of termination) would be appropriate in some circumstances and can be considered.
- As with many other agreements, the company should retain the right to immediately terminate the SDK agreement if the developer breaches the agreement or declares bankruptcy.

Summary

Because an SDK permits a developer to interface with a company's software and hardware, care should be taken to ensure that the company has the appropriate protections in place to protect against the associated risks and reduce its liability. Understanding the key considerations discussed in this chapter and including the foregoing essential terms in your company's SDK agreements can protect the company from the inherent risks associated with SDKs and the liability that could potentially arise.

Chapter 13

Key Issues and Guiding Principles for Negotiating a Software License or OEM Agreement

CHECKLIST

Initial Matters

- ☐ Customer executes mutual NDA
- ☐ NDA only addresses confidentiality
- ☐ Use company's standard license agreement
- ☐ Give customer proposed agreement early
- ☐ Customer should return redline of agreement before negotiation begins
- ☐ No calls to "talk over the agreement" before redlines
- ☐ Respond to customer's redline with company's redline
- ☐ Accept customer's revisions or propose alternative language
- ☐ Show customer's form to legal counsel
- ☐ No negotiating revisions without legal counsel
- ☐ No discussion of agreement with customer's attorneys
- ☐ Create basic term sheet

- ☐ Never rely on redlines produced by customer or third party
- ☐ Require editable Word documents
- ☐ Be cautious in mailing documents outside company
- ☐ Defer negotiation of open provisions
- ☐ Negotiate telephonically
- ☐ Business and legal teams need e-mail access during negotiations
- ☐ Defer negotiation for questions requiring internal consideration

License/Ownership Scope

- ☐ Nonexclusive
- ☐ Nontransferable
- ☐ Nonsublicensable
- ☐ No irrevocable licenses
- ☐ Required development activities?
- ☐ No access to source code
- ☐ Customer must use standard end-user license agreement
- ☐ Licensee has no ownership interest in company's software

Pricing

- ☐ Commensurate with scope of license granted
- ☐ No single-price broad licenses
- ☐ Specify uses for software in fixed-price license
- ☐ Annual maintenance fee
- ☐ Major new versions and new software products
- ☐ Company retains audit rights

Limitations of Liability

- ☐ Includes exclusion of consequential damages and cap on direct damages
- ☐ If mutual, exclude customer misuse of company software and IP
- ☐ Cap on company's overall damages
- ☐ Breaches of confidentiality and willful misconduct exclusions

Warranties

- ☐ Requires material compliance with company's documentation
- ☐ No longer than ninety days
- ☐ Exclusive remedy for breach is repair
- ☐ Implied warranties
- ☐ Written warranties not included in agreement

Support and Maintenance

- ☐ Support vs. maintenance
- ☐ No new versions or new functionality in support
- ☐ No material alterations to standard support program
- ☐ Priced annually
- ☐ Automatic renewal of support term
- ☐ No commitment to support after five years
- ☐ No agreements to provide "free" professional services
- ☐ Initial fixed fees become "then current rates"

Payment

- ☐ Based on objective and easily identifiable event
- ☐ Testing/acceptance language reviewed by legal counsel
- ☐ License fees not subject to refund
- ☐ Monthly invoices
- ☐ No fixed-fee arrangements

Term and Termination

- ☐ Consistency between license type and term of support
- ☐ Initial term with automatic year-to-year renewal
- ☐ Licenses immediately terminate
- ☐ Licenses to end users do not terminate with customer agreement
- ☐ Misuse terminates perpetual license
- ☐ Opportunity to cure before termination for cause

Infringement Indemnification

- ☐ Company liability unlimited
- ☐ Legal counsel drafts indemnification
- ☐ Company controls defense/settlement
- ☐ "Standing alone"
- ☐ Approved list of countries/jurisdictions

Key Issues and Guiding Principles

This chapter discusses the different topics and issues that should be considered in negotiating a customer (i.e., outbound) software license or original equipment manufacturer (OEM) agreement (compare Chapter 2, which focused on "in-bound" agreements where the vendor is licensing a customer a software application). The summary is designed to minimize the time required to negotiate these types of agreements, make the process proceed more smoothly, and ensure your company's business and legal objectives are achieved.

Initial Matters

- The customer should execute a mutual nondisclosure/confidentiality agreement (NDA) before engaging in any substantive discussions.
- The NDA should be mutual (protecting both parties' confidential information).
- Be sure the NDA does not address matters other than confidentiality (e.g., no provisions regarding intellectual property ownership, license rights, representations and warranties, indemnification).
- Insist on using the company's standard license agreement. This will result in a much better agreement and reduce the time and expense needed to conclude. The company's agreement will likely be specifically designed for the transactions the company is contemplating, while the customer's agreement will almost certainly be a generic form agreement that may have little relation to the engagement under consideration. Much time can be wasted trying to revise a customer's agreement to make it fit the intended engagement. Explain to the customer that it is the company's software/intellectual property that is being licensed and that the company's agreement has been specifically drafted to provide a balanced approach to the protection of both parties' interests.
- The company should endeavor to give its proposed agreement to the customer as early as possible in the process.
- Before any substantive negotiations of the legal terms, the customer should be required to return a redline (using Microsoft Word track changes) of the agreement. Customer revisions should be reviewed by legal counsel before negotiations commence. Avoid scheduling calls with the customer to "talk over the agreement" without first requiring a complete redline of the customer's proposed revisions.
- In general, the company's process should be to provide its form agreement, require the customer to respond with a complete redline, and then provide the company's own proposed redline in response to the customer's redline before scheduling any substantive legal discussions. This approach forces the customer to focus on the agreement and commit to providing a complete set of comments. The company can then assess those comments and frequently accept many of the revisions or proposed alternate language. By going through these steps before scheduling substantive negotiations, the company can frequently narrow the issues to only a handful of items.

- If the company must use a customer's form, it should provide the form to legal counsel for review prior to any negotiations concerning or making any revisions to the agreement.
- Do not negotiate revisions to any agreement (other than purely economic/commercial terms) without advice and input of legal counsel.
- For maximum flexibility, it is preferable for the business and legal terms to be negotiated contemporaneously (e.g., resolution of legal term negotiations may impact the company cost, risk, and pricing).
- Do not negotiate or discuss the agreement or relationship with the customer's lawyers unless the company's counsel is present and participating in the discussion.
- Develop a basic term sheet (e.g., term of license, product being licensed, proposed/licensed uses, fee structure, unique terms) for the deal as early as possible in the relationship. The term sheet can guide legal counsel in ensuring the agreement accurately reflects the proposed transaction.
- *Never* rely on a redline prepared by a customer or other third party. Whether intentionally or unintentionally, redlines provided by others are frequently inaccurate. It should be the company practice to always create its own redlines of revised documents.
- Some customers will insist on providing locked/password-protected documents or PDFs of their proposed changes. The company should gently remind the customer that these approaches will greatly slow down the process and are unworkable. Each side should provide editable Word versions of the documents. Without an editable Word version of the document, the company cannot create its own redlines to confirm the changes the customer has made.
- Always check documents before sending them outside the company to ensure they do not include any internal comments or changes that should not be shared with the customer.
- If the customer returns a redline with many provisions marked "open" for further discussion on their side, the company should defer scheduling any negotiation calls until the customer provides a complete redline of its comments.
- Technology licenses are generally negotiated telephonically. In-person meetings are rarely as productive as teleconferences, but can frequently cause parties to make concessions they might not otherwise make because they are "there" and want to get the deal done. In general, scheduling a series of one to two-hour calls to review contract changes and discuss issues is far more productive than scheduling a single, in-person meeting.
- During negotiation calls, the company business and legal teams will likely be in different locations, but should have access at all times to e-mail to raise internal issues or ask questions of the other members of the team.
- If, during a negotiation call, an issue arises that the company believes should be discussed internally, the company can either defer the issue or ask for a brief break in the call so that it can discuss the issue privately.

Scope of License/Ownership

- Licenses should generally be nonexclusive, nontransferable, and nonsublicensable, except to the extent expressly permitted by the agreement (e.g., an OEM would have the right to distribute to end users or other customers).
- Generally, reject requests from customers to grant "irrevocable" licenses. The company should be able to terminate the license if the customer misuses the company's intellectual property or fails to pay the agreed-upon fees.
- Consider whether any development activities are required by the company and/or OEM to integrate the company software into OEM software/products. License rights should be consistent with any development activity (e.g., the company should own all intellectual property that it creates).
- Only object code (the machine-readable version of the software) is licensed. Access to company's source code is forbidden, except in very narrow circumstances (e.g., the company generally ceases to support the product).
- If the customer will be redistributing the company's software to others (e.g., in an OEM arrangement), the customer must be required to include the company's standard end-user license agreement with delivery of software or products to the OEM's customer and ensure that the customer is bound by the company's end-user license agreement. Other arrangements can be made, but it is key to ensuring the company is contractually protected from claims by end users.
- The agreement should make clear the licensee will have no ownership interest in the company's software or any modifications that the company makes to the software. Avoid all language that provides for the transfer of intellectual property rights from the company to the customer. A simple clear statement that, apart from the express license being granted, all other rights are reserved to the company and that the agreement will not be construed as a sale of any intellectual property rights.

Pricing

- Pricing should be commensurate with the scope of the license granted (e.g., license fee for perpetual term will be more than for time-limited term, broad distribution rights cost more than narrow rights, the specific types of applications in which the company software may be used should be clearly identified). Avoid single-price, broad licenses (e.g., a one-time perpetual license for "all applications").
- If a fixed-price license is considered (e.g., does not vary based on number of users, installations, sites), the specific uses to which the software may be put should be clearly specified.

- In addition to the license fees, the customer must pay an annual maintenance fee to receive support and updates.
- Major new versions and new software products are not included in the standard support and maintenance. These may be licensed for an additional fee.

Audit Rights

- Unless the agreement involves a fixed fee, the company should have the right to periodically audit the customer to confirm compliance with the license agreement and proper calculation of fees due to the company. In the event the customer has substantially underpaid the fees due (e.g., greater than 10%), the cost of the audit should shift to the customer.
- Audit rights are key to ensuring the customer accurately reports and pays the fees due to the company.

Limitations of Liability

- There are two parts to almost every limitation of liability: (i) disclaimer/exclusion of consequential damages (e.g., lost profits) and (ii) cap on direct damages (e.g., one year of fees paid by customer). The company should require a limitation of liability in every contract.
- The company could agree to make the limitation of liability mutual, protecting both parties, but the company must ensure the customer's misuse of company software and other intellectual property is excluded from all limitations and exclusions of liability.
- In general, the company's overall damages should be capped at the fees paid by the customer to the company in the three to six months immediately preceding the event giving rise to liability. This can be increased to twelve months, but further increases are strongly disfavored.
- Limitations of liability are commonplace in software licensing transactions and should be insisted on in all of the company's agreements.
- Except for the company's indemnity obligation, the company should not accept any other exclusions from its limitation of liability. If pressed, the company may agree to exclude breaches of confidentiality and willful misconduct. Other revisions must be approved by legal counsel.

Warranties

- Warranty should be limited to material compliance with the company documentation and no longer than ninety days. Paid support and maintenance provide protection following the warranty period.

- Exclusive remedy for breach of the warranty should be limited to the company repairing or replacing the software.
- Certain statutory warranties may be "implied" and should therefore be specifically disclaimed/excluded. Any written warranties or statements not in the agreement (e.g., marketing materials) should also be disclaimed and excluded.
- Avoid numerous additional warranties.

Support and Maintenance; Professional Service Rates

- Typically "support" is help desk support and error correction/bug fixes; "maintenance" is providing periodic releases and updates to the software.
- Make clear the customer is not entitled to new versions or new functionality as part of standard support. Be wary of customer requests to rewrite or expand the company's definitions of bug fixes and updates. These revisions are typically designed to give the customer new versions or other functionality for which the company would otherwise charge an additional fee.
- Unless the transaction is very substantial, the customer must understand that the company cannot materially alter its standard support program. In particular, the company cannot commit to support service level requirements that are different from its normal support metrics. Service levels should be framed as "goals" or indicate the company will use commercially reasonable efforts to achieve those levels, typically priced annually, include automatic renewal of the support term, if possible.
- Avoid committing to support the software indefinitely. The company should not commit to provide support for more than five years from the initial date of the license.
- Avoid support commitments that could require the company to provide "free" professional services. If the customer calls with a support problem that turns out to be caused by the customer's own products, the company should have the ability to charge for its time.
- While rates can vary based on services provided, or nature of the software, support and maintenance are often priced based on a percentage (e.g., 15%–25%) of the license fee. Avoid agreeing to cap support fees for more than three to five years.
- Some customers may ask the company to fix the company's time and materials rates for professional services. The company may agree not to increase its fees for six months to a year, but should generally avoid longer commitments. After the initial period in which the fees are fixed, they become the company's "then current rates."

Payment

- Payment of license fees should not be contingent or uncertain. Payment should be based on an objective easily identifiable occurrence (e.g., delivery of the software to the customer).
- Any testing and acceptance language insisted on by the customer should be carefully worded, strictly time-limited, and reviewed by legal counsel. Avoid keying acceptance or the triggering of fees on events within the sole control of the customer. Acceptance and payment events should be keyed to objective milestones or passage of time.
- License fees should not be subject to refund, except for failed acceptance testing.
- Unless agreed differently, payment for services (e.g., installation, customization, and implementation) is typically based on monthly invoices from the company based on time and materials basis. Avoid fixed-fee arrangements if possible.
- It is often important to explain to the customer that due to certain revenue recognition rules, license fees and other payments cannot be contingent or subject to refund. The company must ensure it will be fairly compensated for its software and any services provided.

Term and Termination

- The term (length) of the agreement must be consistent with license type and the term of support. For example, a perpetual license may be granted with a support term that only lasts for several years. Support terms are typically written with an initial term (e.g., one to three years) with automatic year-to-year renewal thereafter.
- Generally, upon termination of the agreement, all licenses immediately terminate. Provisions can be made for perpetual license continuing even if support is terminated. In any event, language is typically included making clear that licenses to end users do not terminate on termination of the customer's agreement.
- Even a perpetual license is subject to termination for misuse of the software or breach of any license restrictions.
- Each party should have notice and opportunity to cure before termination for cause (except breach of confidentiality, misuse of the software, or bankruptcy, which may result in immediate termination).

Infringement Indemnification

- The company could indemnify the customer if it is sued by a third party claiming that the company software, standing alone, infringes the third party's patent, copyright, or trade secret rights.

- Provisions must be carefully drafted by legal counsel, as the company's liability for infringement is typically unlimited.
- In the event of a claim, the company controls the defense and settlement of the claim and the customer must cooperate. The indemnity language may also require the customer to avoid making any admissions regarding the litigation unless consented to by the company or required in sworn discovery responses.
- The company should only indemnify for claims arising out of the company's software "standing alone"—e.g., not combined with third-party products or software, including any intellectual property of the licensee.

Summary

The summary provided in this chapter highlights different topics and issues that a company should consider in negotiating a customer software license or OEM agreement. The checklist and summary provided should be used to minimize the time required to negotiate these types of agreements, making the process more seamless and protecting the company's business and legal objectives.

Chapter 14

Drafting OEM Agreements (When the Company Is the OEM)

CHECKLIST

Scope of Engagement

- ☐ Agreement covers all relevant activities
- ☐ NDA prior to substantive discussions
- ☐ Describe coordination efforts
- ☐ Who will contact customers?
- ☐ "Private label"?

Customer Terms

- ☐ Controlling liability
- ☐ Appropriate license
- ☐ Terms and conditions
- ☐ Absolve of liability
- ☐ Appropriate drafting

Territory

- ☐ Geographic restrictions
- ☐ Clear identification of restrictions

Hardware Products

- ☐ How order will be placed
- ☐ How order will be filled
- ☐ Timing of orders
- ☐ Return procedures
- ☐ Warranty claims
- ☐ Nonbinding order projections
- ☐ Price reduction incentives

Exclusivity

- ☐ Address exclusivity explicitly
- ☐ Specific revenue commitments
- ☐ Remedy is not breach of contract
- ☐ Broad termination rights

Supplier Product Changes

- ☐ Obligation to coordinate with company

Support and Training

- ☐ Train personnel adequately
- ☐ Ensure supplier cooperation
- ☐ Specify service levels

Confidentiality

- ☐ Strong clause in agreement

IP Issues

- ☐ Standard commercial licenses
- ☐ Terms and conditions
- ☐ Returning products
- ☐ Reserve rights explicitly
- ☐ No transfer or assignment

Warranties/Disclaimers

- ☐ Authority to enter agreement
- ☐ Noninfringement

- ☐ Free of known defects
- ☐ Unaware of litigation
- ☐ Freedom from viruses
- ☐ Freedom from disabling code
- ☐ Compliance with relevant codes

Limitations of Liability

- ☐ Direct and consequential damages
- ☐ Last three months of fees paid
- ☐ Injury to persons

Indemnification

- ☐ IP infringement claims
- ☐ Supplier violation of applicable law
- ☐ Products liability
- ☐ IP infringement claims caused by OEM
- ☐ Limited jurisdictions
- ☐ Term and termination

Specific Initial Term

- ☐ Agreement to renew company's option
- ☐ Automatic renewal
- ☐ Company right to terminate without cause
- ☐ Revenue commitments?
- ☐ Breach of agreement
- ☐ Sell-off period
- ☐ Continue to support existing custom

Key Issues and Guiding Principles

This chapter discusses the different topics and issues that should be considered in entering into original equipment manufacturer (OEM) arrangements where the company will be the OEM (i.e., the company will take the software and hardware products of a supplier, combine those products with the company's own products, and then sell the combination). These types of transactions can take several forms. In some engagements, the supplier's assistance may be required to integrate its products with the company's products. In other engagements, the company may want to "private label" the combined product (i.e., market and sell the combined product without identifying the supplier). Still other engagements may require close interaction of the parties in rendering support and warranty service to customers. Depending on the type of relationship, more detailed agreements may be required to fully address the parties' respective integration and support obligations.

The summary provided in this chapter is designed to minimize the time required to negotiate these types of agreements, make the process proceed more smoothly, and ensure the company's business and legal objectives are achieved.

Determine the Scope of the Engagement

- As discussed in several other chapters, these types of relationships can take several forms. The first step is to determine the scope of the engagement and to ensure the agreement covers all relevant activities.
- Depending on the nature of the information to be shared between the parties, a nondisclosure agreement should be entered into prior to commencing substantive discussions.
- If substantial coordination will be required to integrate the supplier's products with the company's products, a detailed description of the coordination efforts should be attached to the OEM agreement.
- Another preliminary matter is to determine who will have contact with customers for support issues. The company can provide "first-level" support directly to customers, with the supplier providing "second-level" support to the company's own support personnel. In private-label arrangements, the company should have complete control over all contacts with customers.
- Decide whether the engagement will be "private label" or if the supplier will be identified in some way in connection with the combined product.

Customer Terms

- Appropriately drafted customer terms will protect both the company and the supplier from liability and minimize the potential for customer claims. This is a key protection that is frequently overlooked or not addressed properly.
- Since both parties will be providing their products to the ultimate customer, both parties will be concerned with controlling their liability to those customers. This control is achieved by providing the combined product to customers under an appropriate license or other terms and conditions, to be drafted by the company. It is common in the industry for the supplier to request that those terms and conditions protect its intellectual property and absolve it of

liability directly to the customer. This can be achieved by the parties specifically agreeing on the terms to be presented to customers. Another common approach is to require the OEM, in this case the company, simply to commit to ensuring the customer terms will protect the supplier from liability, disclaim warranties, and protect its intellectual property.

Territory

- In many OEM agreements, the supplier may request restrictions on the geographic territory in which the combined product can be sold. In general, it is best to seek the ability to sell to the entire world, but this may not be possible in all engagements. The point is to ensure that if there are territorial restrictions, those restrictions are clearly identified in the contract.

Hardware Products

- If the supplier will be furnishing hardware products, the agreement should address how the order will be placed and filled, the timing of those orders, return procedures, and handling of warranty claims.
- To ensure the supplier can satisfy the company's requirements, it may be appropriate to include the ability of the company to provide the supplier with nonbinding projections of future order volumes.
- If delivery timing is critical, incentives in the form of price reductions should be included to ensure the supplier delivers on time (e.g., the sales price might be decreased by 3% for each week the supplier is late in delivering its products).

Exclusivity

- If the supplier is to be exclusive to the company (i.e., the supplier can furnish its products only to the company), the supplier may require specific revenue commitments from the company. The remedy for failure to achieve those commitments should not, however, be breach of contract, but merely loss of exclusivity.
- If the supplier is seeking exclusivity from the company (i.e., that the company will not buy similar products from another supplier), the contract must have appropriate and broad termination rights for supplier performance issues, product defects, infringement claims, and other similar matters.

Supplier Product Changes

- If the supplier's product will be closely integrated into the company's products, the agreement should impose obligations on the supplier to coordinate with the company in the event the supplier contemplates any changes to its products that could impact that integration. In most instances, the parties will want to agree on joint testing protocols to ensure the combined product works properly.

Support and Training

- The agreement should require the supplier to train company personnel adequately about the supplier's products, ensure the supplier cooperates with the company in rendering support and processing warranty claims, etc. If support response time is critical, the agreement should specify service levels for the supplier's support obligations. The contract may also address the possibility of doing "warm handoffs" from the company to the supplier of certain customers to handle highly unique or complex issues.

Confidentiality

- Since the parties will likely be exchanging product information, marketing plans, customer names, and other sensitive information, the agreement should include a strong confidentiality clause.

Intellectual Property Issues

- If the company will be providing any software or other products to the other party for purposes of joint support, product integration, or other activities, the products should be provided under their standard commercial licenses and/or product terms and conditions. Since the products are not being provided as the result of a normal license or purchase transaction, the agreement should make clear the products are provided on an as-is basis. The products should be returned on expiration or termination of the agreement or at any time on the company's request.
- The agreement should contain a clear statement that each party reserves all rights in its intellectual property and that neither party is transferring or assigning any rights in its intellectual property.

Warranties and Disclaimers

- At a minimum, the agreement should contain a range of warranties from the supplier, including the following:
  - Authority to enter into the agreement
  - Noninfringement
  - The products are free of known defects
  - The supplier is unaware of any litigation, including products liability claims, relating to its products
  - Freedom from viruses and disabling code
  - Compliance with all fire, electrical, environmental, health and safety, and other relevant codes, regulations, and laws
- Unless the agreement includes specific revenue commitments, a disclaimer should be added making clear that neither party is guaranteeing the other party will derive any specific revenue from the relationship.

Limitations of Liability

- Apart from liability relating to indemnity obligations, breach of confidentiality, or a party's misappropriation of the other party's intellectual property, the agreement should disclaim essentially all other liability for both direct and consequential damages. It is common to limit liability to several (e.g., the last three) months of fees paid, if any.
- If consumer products are involved, another exclusion from all limitations of liability should be added for injury to persons.

Indemnification

- The supplier should be required to indemnify the company from intellectual property infringement claims (unless a claim arises from the company's combination of the supplier's product with something the supplier has not provided), the supplier's violation of applicable law, and products liability claims (particularly if hardware is being supplied). If the supplier insists, it is common to require the OEM, in this case the company, to provide an indemnity from any intellectual property infringement claims that it causes.
- If the supplier seeks to limit the jurisdictions in which it will provide an intellectual property indemnity, make sure those jurisdictions include all areas of the territory in which the company has been licensed to sell the combined product.

Term and Termination

- The agreement should contain a specific initial term. Since the supply of the products may be critical to the company's marketing plans, the agreement should renew at the company's option (but not the supplier's option) for an additional two to three years. Thereafter, the agreement would typically automatically renew for additional one-year terms unless either party gives notice of its intent not to renew.
- Unless business reasons dictate otherwise, the company should have the right to terminate the agreement for convenience at any time, without cause, on written notice to the supplier (e.g., thirty days' prior notice). This may not be possible where revenue commitments are required.
- Either party should have the right to terminate the agreement if the other party breaches the agreement.
- In the event of a termination, it is critical to include a "sell-off period" during which the company can sell off the stock of combined products on hand. This period can range from six months to a year.
- Regardless of the reason for termination, the company must ensure it has the right to continue to support the existing customer.

Summary

OEM arrangements where the company is the OEM can be sensitive in terms of protecting all of the company's interests. The summary included in this chapter should reduce the difficulty of negotiating these agreements, making the process progress quickly and ensuring the company's business and legal objectives are agreed.

Chapter 15

Original Equipment Manufacturer (OEM) Agreements

CHECKLIST

Key Contractual Concerns from the Perspectives of Both Parties
Scope of Use of Supplier Technology

- ☐ Supplier technology may not be exploited separately from the OEM product

Distributors and Subdistributors

- ☐ OEM's right to use distributors and subdistributors
- ☐ OEM liability for the actions of distributors and subdistributors

Private Labeling

- ☐ OEM's right to private label
- ☐ Supplier's right to ensure its intellectual property rights are protected

Source Code Access

- ☐ OEM access to source code
- ☐ OEM's liability for functionality and other issues resulting from modifications it makes to source code

Open-Source Software

- ☐ Whether supplier can include open source in its products
- ☐ Viral licenses
- ☐ Protection of OEM proprietary software

Territory

- ☐ OEM's right to sell its products in geographic or industry-specific areas

Support Obligations

- ☐ First-level end-user support
- ☐ Second-level support to OEM

Training

- ☐ Supplier training of OEM personnel

End-User/Customer Liability

- ☐ Protecting supplier from claims and damages from end users
- ☐ Requiring the OEM to protect the supplier
- ☐ Pass-through terms
- ☐ OEM indemnity for failure to protect supplier

Fees

- ☐ All you can eat
- ☐ Per unit
- ☐ Percentage of OEM product fees

Audit

- ☐ Confirmation of calculation and payment of fees

Expansion of Scope

- ☐ OEM's rights to expand the scope of the engagement

Professional Services

- ☐ Supplier professional services to modify and integrate its product
- ☐ Professional service terms

Sell-Off Period

- ☐ OEM's right to sell off products on termination of the agreement
- ☐ Ongoing support

Other Issues

- ☐ Term and termination
- ☐ Warranties
- ☐ Indemnities
  - – Indemnity for unauthorized warranties
- ☐ Limitation of liability
- ☐ Confidentiality
- ☐ Export/import
- ☐ Acquisitions by the federal government

Overview

An Original Equipment Manufacturer ("OEM") Agreement involves an arrangement where one party, sometimes called the *supplier,* furnishes technology to the other party, the OEM, for combination with other technology for redistribution as a complete product to customers, end users, and so on. The OEM essentially takes components provided by one or more suppliers, combines them with technology of the vendor, and creates a finished product. The OEM's activity is sometimes referred to as *adding value* to the components supplied by the supplier(s).

OEM agreements are written from either the perspective of the OEM, the party making the distribution of the finished product, or the supplier, the party furnishing one or more pieces for incorporation into the finished product. The parties frequently have divergent interests. The OEM wants broad distribution rights, strong protections from claims of intellectual property infringement relating to the supplier's technology, the right to potentially distribute the finished product under a private label arrangement (i.e., an arrangement in which the finished product appears to have been entirely created by the OEM), clear support obligations, and so on. In contrast, the supplier may be concerned about controlling liability for claims from end users or customers purchasing the finished product, protecting its intellectual property from misuse by end users and customers, limiting its responsibility for intellectual property infringement claims when its technology is combined with other technology, and so on.

As an aside, the difference between an OEM agreement and a reseller agreement should be noted. In a reseller agreement, the supplier's technology is sold or licensed by the reseller as is (i.e., not as part of a larger product, but as a standalone product). In an OEM agreement, the supplier's technology is never sold as a stand-alone product, but only as part of a larger product. In fact, it would be a breach of the OEM agreement to distribute the supplier's technology in a stand-alone fashion.

Key Contracting Concerns from the Perspectives of Both Parties

Since a business may find itself serving in either the position of a supplier or OEM, the following provides a discussion of the key issues presented to both parties in these engagements. See Chapter 14 for issues focusing on the perspective of the OEM.

- **Scope of use of supplier technology.** As a foundational issue, the agreement should define exactly how the supplier technology may be used. In most cases, that means identifying the specific OEM products or lines of products into which the supplier technology may be incorporated.

  As noted above, the supplier will want to make clear that its technology can never be sold or licensed as a stand-alone product, but always as part of the overall OEM product; otherwise, the engagement would be nothing more than a standard reseller arrangement.
- **Distributors and subdistributors.** In most OEM engagements, the OEM will require the use of distributors and subdistributors to get their products to market. The OEM agreement should spell out how these third parties may be used and the OEM's liability for their actions.
- **Private labeling.** Depending on the business goals of the relationship, the OEM may want the right to private label the final product. This means the final product will be sold as if entirely created by the OEM, with no mention of the supplier(s).

  In private label arrangements, the supplier will want to require the OEM to nonetheless include all notices and references to the supplier as are necessary under the law to protect its intellectual property rights (e.g., copyright notices in the technology identifying the supplier must be retained).
- **Source code access.** Apart from requiring the supplier to potentially escrow its source code to protect the OEM in the event the supplier ceases to do business or files for bankruptcy protection, the agreement should address whether the OEM needs access to the supplier's source code from the outset (e.g., to integrate it into the OEM's product, create interfaces), and if so, the uses that the OEM may make of the source code.

  If the OEM has access to the source code, the supplier will be concerned with operational issues, bugs, and other problems that may result from the OEM's modifications of the source code. Suppliers should disclaim liability for those issues (both for breaches of warranties and for intellectual property infringement claims resulting from those modifications) and, potentially, charging on a time and materials basis to assist in resolving those issues. Such work would be in addition to the OEM's standard support obligations.

- **Open-source software.** Open-source software is software provided in source code form that is made available under specialized licenses permitting their free and easy copying, modification, and distribution. The software is generally provided without license fees. Certain open-source software is provided under license agreements that contain a "viral" aspect. That is, if open-source software governed by a viral license is combined with proprietary software and redistributed, the viral license may require, among other things, that the combined product be provided without charge and that the source code of the combined product be disclosed. The most common viral license is the General Public License (http://www.gnu.org). In the context of an OEM engagement, if the supplier furnishes software to the OEM containing open-source software governed by a viral license, the result could be catastrophic for the OEM. Its valuable proprietary software that it combines with the supplier software could become subject to the viral license, requiring the OEM to essentially give away its software and disclose its source code to end users. As a result, all OEM agreements involving software should contain a strict prohibition on open-source software governed by these types of viral licenses. In most instances, to mitigate risk, OEMs prohibit any open-source software in the supplier's product.
- **Territory.** The agreement should specify the territory in which the OEM may sell or license its products containing the supplier technology. The territory may be geographic (e.g., North America) or industry-specific (e.g., healthcare or financial services).
- **Support obligations.** The agreement should define the parties' respective obligations with regard to support. Generally, the OEM is responsible for first-level support (i.e., support involving direct contact with end users or customers), while the supplier provides second-level support to the OEM's support personnel. The agreement may define how the parties will integrate their support processes and personnel, support service levels, the OEM's right to updates and enhancements to the supplier's technology, and so on. The OEM's concerns about the supplier's support obligations are similar to those in any license agreement.
- **Training.** In addition to support obligations, the OEM agreement may include language requiring the supplier to provide certain training to OEM personnel to ensure they have sufficient knowledge to operate and use the supplier's technology.
- **End-user/customer liability.** Potential liability of the supplier to end users and customers is a thorny problem. Since the supplier generally has no direct contract with the end user or customer (i.e., no "contractual privity"), it cannot limit its liability as it normally would through an end-user agreement.

The supplier is concerned with the possibility that the OEM product malfunctions or otherwise causes damage to the end user or customer; the end user or customer sues the OEM for damages, conducts discovery, and finds out that portions of the product were furnished by supplier; and sues the supplier. In such a case, the OEM will likely have fully protected itself by virtue of its direct contract with the end user or customer, but the supplier has no such protection. The other concern suppliers have with regard to end users and customers is the protection of its intellectual property (i.e., they need limits imposed on the end users and customers so that they do not misappropriate or misuse the supplier's intellectual property). There are several ways to address these concerns:

- Include language in the OEM agreement requiring the OEM to include in its end-user agreements (i) a complete disclaimer of liability and warranties for its licensors and suppliers and (ii) protections for the supplier's intellectual property consistent with those in the OEM agreement. In this approach, the supplier is relying on the OEM to draft appropriate contract language to protect the supplier.
- Require the OEM to include the protections above and provide a copy of the OEM's end-user agreement for the supplier's review and approval prior to any sale or license of the OEM product. In this approach, the supplier can assure itself that the OEM has properly protected it.
- Require the OEM to pass through to each end user certain terms and conditions supplied by the supplier. In this approach, the supplier has direct control over the terms presented to the end users. This approach, however, prevents the OEM product from being offered on a private-label basis because the end user is put on notice of the supplier through the supplier's required pass-through terms.

  In all of the approaches previously discussed, suppliers generally require the OEM to indemnify and hold them harmless from the OEM's failure to fulfill its contractual obligations to protect the supplier.

- **Fees.** Fees payable under OEM agreements can be calculated in a variety of ways. The OEM may purchase an "all you can eat" license permitting them to deploy an unlimited number of copies of the supplier's technology in the OEM's products. In other cases, the OEM may be required to pay the supplier a per-unit fee for each OEM product distributed containing the supplier's technology. In still other cases, the supplier's fee may be based on a percentage of the fee charged by the OEM for the finished product.

  Fees payable by the OEM are frequently paid on a monthly or quarterly basis, with true-ups to reflect product returns, uncollected fees, and other similar adjustments.
- **Audit rights.** If fees are calculated on a per-unit basis or as a percentage of the OEM's product price, suppliers will generally require rights to audit the OEM's records to ensure all fees have been calculated correctly and paid.

- **Expansion of scope.** Depending on the engagement, it may be appropriate for the parties to discuss and prenegotiate terms for expansion of the scope of use of the supplier's technology (e.g., for other OEM products, for other territories).
- **Professional services.** In many instances, the OEM may need the supplier to render certain professional services to assist in modifying its products or integrating them with the OEM's products. The agreement should include appropriate professional service terms (e.g., statements of work, project management and staffing, change orders, professional service rates) to govern those services.
- **Sell-off period.** It is common in OEM agreements to include a sell-off period on termination of the agreement. That is, the OEM will have an opportunity for a defined period of time following termination to continue to sell or license the OEM products containing the supplier's technology and also an ongoing right to support end users who have previously purchased the OEM's products. In such cases, the terms of the agreement will continue to apply during the sell-off period, provided that the OEM cannot sell or license the OEM products to any new end users.
  - Sell-off periods frequently range from six to twelve months.
  - Suppliers may limit or preclude any sell-off period if termination results from the OEM's breach of the agreement, particularly breaches relating to protection of the supplier's intellectual property or payment of fees due the supplier.
- **Other issues.** Common to any form of technology agreement, the OEM agreement should address the following issues:
  - Term and termination (subject to the sell-off period)
  - Warranties
  - Indemnities
    - Suppliers typically disclaim any liability for warranties made by the OEM relating to the supplier's technology unless the supplier has expressly authorized those warranties in writing. In addition, suppliers generally require the OEM to indemnify them and hold them harmless from any claims and damages resulting from unauthorized warranties.
  - Limitation of liability
  - Confidentiality
  - Export/import
    - While suppliers may cooperate with OEMs in obtaining export or import licenses, OEM agreements generally place the burden of complying with export and import laws on the OEM.
- Acquisitions by the federal government
    - To protect both parties' intellectual property, language should be included that is designed to mitigate any claim by the government that it is obtaining an ownership interest in their intellectual property.

Summary

OEM agreements afford suppliers an additional opportunity to exploit their technology and afford OEMs the ability to leverage technology created by others. While those agreements present many of the same issues presented in standard license and reseller engagements, there are issues relating to intellectual property, liability to end users, scope of use, and other issues found in no other type of technology engagement. Depending on your position in the engagement (i.e., OEM or supplier), you must ensure the agreement addresses your specific concerns.

Chapter 16

Health Insurance Portability and Accountability Act (HIPAA) Compliance

CHECKLIST

HIPAA/HITECH Compliance

- ☐ Health Information Technology for Economic and Clinical Health (HITECH) Act
- ☐ Civil and criminal penalties
- ☐ Expanded definition of business associates (BAs)

Who Are BAs?

- ☐ Working on or behalf of covered entities (CEs)
- ☐ Providing protected health information (PHI) data to CEs
- ☐ Vendors contracting with CEs

Fail to Comply with HIPAA

- ☐ CMPs: $100–$10,000/violation
- ☐ Criminal penalties

- ☐ Mandatory Health and Human Services (HHS) investigation and assessment
- ☐ Civil actions by state attorney generals (AGs)

Security Breach Notification

- ☐ Must notify CEs of unsecured PHI breaches
- ☐ CEs must notify individuals
- ☐ CE may need to notify HHS and local media
- ☐ BAs bear burden to prove reasonable delay in notification
- ☐ Security breaches of unsecured PHI include unauthorized acquisition, access, use, or disclosure of PHI
- ☐ Unsecured PHI is not encrypted or destroyed
- ☐ CEs must notify patients within sixty days after discovery of breach
- ☐ Date of discovery or date breach should have been discovered
- ☐ Information BAs provide to CEs following breach
- ☐ Contractual obligations of BAs to notify on behalf of CEs
- ☐ Compliance with state laws
- ☐ BAs' internal policy for notification
- ☐ Contractual binding of subcontractors

HIPAA Security Rule

- ☐ Administrative, physical, and technical safeguards
- ☐ Specific standards of implementation
- ☐ Gap analysis for shortfalls
- ☐ HHS recommends technical safeguards
- ☐ Subcontractor agreements
- ☐ Information security due diligence questionnaire

Statutory Liability

- ☐ Amending noncompliant BAAs
- ☐ Renegotiate with CEs
- ☐ BAAs increase in complexity
- ☐ Indemnifying CEs
- ☐ Required notification of breach on behalf of CEs
- ☐ Responsibility for costs of breach
- ☐ Draft form amendments to BAAs
- ☐ Minimize negotiation terms not required by law
- ☐ Reflect obligations of BAs, but protect from liability for subcontractor breaches

Additional HIPAA Requirements

- ☐ Comply with minimum necessary standards
- ☐ Use of a limited data set?
- ☐ Ongoing assessment of what is minimum necessary
- ☐ CEs must account to individuals of disclosures from electronic health records (EHRs)
- ☐ Monitor developing HHS advice
- ☐ No direct or indirect remunerations to BAs for EHR or PHI
- ☐ Making recommendations for products or services

Steps for Breach Notification Compliance

- ☐ Analyze existing policies and procedures
- ☐ State breach notification requirements?
- ☐ Designate person to ensure breach investigation and determine if breach occurred
- ☐ Outside legal counsel
- ☐ No unreasonable delay in reporting
- ☐ Impacted individuals identified
- ☐ Impacted individuals reported to CE
- ☐ Employees trained on reporting breaches and handling PHI
- ☐ Sanctions for employees
- ☐ Can BA-controlled PHI be secured?
  - Encrypted
  - Destroyed
- ☐ Amend existing reporting policies
- ☐ Seek outside legal review of amendments
- ☐ Risk prevention and mitigation strategies
- ☐ Decrease risk of breach?
- ☐ Insurance covers costs from breach?

Steps for Security Rule Compliance

- ☐ Perform gap analysis
  - Administrative safeguards
  - Physical safeguards
  - Technical safeguards
- ☐ Make written policies and procedures for each standard above
- ☐ Seek legal review of policies
- ☐ Train employees on requirements

Amendment of BAAs

- ☐ Draft template amendments
- ☐ CE may conduct due diligence of BA
- ☐ Negotiate broad indemnification or cost-allocation provisions
- ☐ Terms in existing service agreements conflict with BAA?
- ☐ Amend subcontractor agreements

Inventory HIPAA-Related Policies

- ☐ Current policies facilitate compliance?
- ☐ Accounting for disclosures made from an EHR?
- ☐ Minimum necessary disclosures/limited data set?
- ☐ Prohibition on sale of EHRs or PHI?
- ☐ Conditions on marketing communications?
- ☐ Training procedures for personnel?
- ☐ Review sanctions for employee violations

Overview

Key Issues and Guiding Principles

The Health Information Technology for Economic and Clinical Health (HITECH) Act, enacted as part of the American Recovery and Reinvestment Act of 2009, marked a fundamental change in the federal government's approach to ensuring compliance with *Health Insurance Portability and Accountability Act* (HIPAA) privacy and security rules.[1] Under the HITECH Act, the federal government, in an effort to strengthen HIPAA, enacted a rigorous enforcement strategy that includes strict privacy and security standards, increased penalties for violations, and expanded federal and state enforcement authority, all of which are directly applicable to Business Associates (BAs).

BAs' obligations and exposure under HIPAA are both contractual (BAAs) and statutory (HIPAA/HITECH). Consequently, in addition to being liable under their business associate agreements (BAAs), BAs will are subject to many of the legal requirements set forth in the HIPAA privacy and security rules, including civil and criminal penalties. Further, the HITECH Act expanded the definition of BAs under HIPAA. Certain vendors of personal health records (PHR) systems and certain data transmission organizations, such as Regional Health Information Organizations, are considered BAs and are subject to HIPAA.

[1] Throughout this chapter, the term HIPAA will be defined to include the provisions of the HITECH Act, unless otherwise specifically noted.

This chapter highlights key provisions in the HITECH Act that apply to BAs and provides a high-level outline of some important steps to aid a BA to achieve HIPAA compliance. This document is intended to provide general, high-level guidance only and is not intended to provide or be a substitute for legal advice. BAs should consult legal counsel to understand their obligations under HIPAA and the HITECH Act.

Who Are BAs?

- BAs generally include entities engaged in certain administrative activities or services for or on behalf of covered entities (CEs), which required access to protected health information (PHI), including claims processing, billing, benefit management, utilization review, management services, and consulting services. A BA creates, receives, maintains, or transmits PHI on behalf of a CE or another BA.
- Organizations should keep in mind that the definition of a BA also includes the following organizations:
  - Organizations providing PHI data transmission to CEs such as health information exchange organizations, regional health information organizations, and e-prescribing gateways.
  - Vendors contracting with CEs to provide PHR systems to patients.[2]
- This definition of what constitutes a BA subjects many vendor and service provider organizations to the HIPAA laws governing the privacy of medical or health information.
- BAs can also include developers of health-related mobile apps and personal health trackers and devices. The Federal Trade Commission (FTC), in conjunction with OCR, the HHS Office of National Coordinator for Health Information Technology (ONC), and the Food and Drug Administration (FDA), created a web-based tool to help developers of health-related mobile apps and similar technology understand what federal laws and regulations might apply to them. The guidance tool asks developers a series of questions about the nature of their app, including about its function, the data it collects, and the services it provides to users. Based on a developer's answers to those questions, the guidance tool points the app developer toward detailed information about certain federal laws that might apply. These include the FTC Act, the FTC's Health Breach Notification Rule, the Health Insurance Portability and Accountability Act (HIPAA) Rules, and the Federal Food, Drug and Cosmetics (FD&C) Act. The tool can be found here (https://www.ftc.gov/tips-advice/business-center/guidance/mobile-health-apps-interactive-tool).

[2] Vendors that provide PHR systems, but do not do so on behalf of CEs, will be subject to security breach notification under the HITECH Act, which will be enforced by the FTC, rather than HHS.

What Can Happen to BAs That Fail to Comply with HIPAA?

- BAs are subject to mandatory periodic audits by the Office for Civil Rights, the US Department of Health and Human Services (HHS) agency responsible for monitoring and enforcing the HIPAA privacy and security rules. BAs found to be noncompliant will be considered to be in violation of the law and subject to the following:
  - civil monetary penalties of between $100 and $10,000 per violation, with maximum penalties of $1.5 million per calendar year;
  - criminal penalties for HIPAA violations; and/or
  - a mandatory HHS investigation and assessment of civil monetary penalties (in cases of willful HIPAA violations).
- Civil actions brought by state attorneys general for HIPAA violations that involve residents in their individual states.

BA Requirements under the Security Breach Notification Requirements

- BAs must notify the CEs with whom they contract of any breaches of "unsecured PHI" and, to the extent possible, identify the individuals whose information was compromised. Upon receiving notice of a reportable "security breach," the CEs have the responsibility to notify the individuals whose information has been breached. In some circumstances, the CEs must also provide notice to HHS and to local media. Notification must take place without unreasonable delay or no later than sixty calendar days from discovery, as required by law. BAs will bear the burden of proof for demonstrating that any delay in notifying the CEs of a security breach was reasonable. Except as required by law enforcement officials, BAs must notify the CEs no later than sixty calendar days from the date of discovery.
- The HITECH Act defines security breach to include the unauthorized acquisition, access, use, or disclosure of PHI that compromises the security or privacy of such information, with certain exceptions for inadvertent acquisition, access, or use of PHI by employees and agents. It is important to note that unless an exception applies, inappropriate acquisition, access, or use of unsecured PHI by employees is considered a reportable security breach.
- Security breaches apply only to unsecured PHI. HHS has issued a guidance document (HHS Guidance) defining the technologies and methodologies to secure PHI, thus rendering the data unusable, unreadable, or indecipherable.

Essentially, PHI must be either encrypted or destroyed per the HHS Guidance to be considered secured. If PHI is secured in accordance with the HHS Guidance, then unauthorized access to, use or disclosure of such information will not trigger the security breach notification requirements. However, such breaches may still be subject to state law notification requirements.

- CEs are required to notify patients without unreasonable delay and in no case later than sixty calendar days after discovery of the breach. The date of "discovery" may not necessarily be the date of actual discovery, but rather, the date that one should have discovered the breach using reasonable measures. Therefore, CEs and BAs should ensure that reasonable measures are in place to catch potential security breaches, as well as to train employees properly to be able to spot these potential breaches. BAs must timely report security breaches to CEs to enable them to notify the individuals within this deadline. It is likely that CEs will amend BAAs to impose tight deadlines on BAs to report security breaches to the CEs.
- BAs are required to include certain information about affected individuals to the CEs to enable the CEs to properly notify affected individuals. The notification should include a brief description of the incident, including the date of the breach and date it was discovered, and the type of unsecured PHI that was breached. CEs will likely require BAs to include additional information regarding the breach, as CEs may need additional information to satisfy their requirements for providing notification to the affected individuals. In some circumstances, CEs may look to contractually obligate BAs who are the subject of a security breach to make the required notifications on behalf of the CEs. The BAs will need to ensure their notification is compliant with HIPAA requirements.
- HIPAA does not preempt more stringent state laws. Essentially, this means that BAs subject to state security breach notification laws must continue to comply with those laws. BAs should consult with legal counsel for assistance with defining these obligations and conducting necessary preemption analysis.
- BAs must develop policies and internal procedures to ensure a coordinated system for internal reporting of breaches of unsecured PHI, prompt internal investigation of alleged breaches, and reporting to the CEs with whom they contract.
- BAs that use subcontractors will have to ensure that they contractually bind their subcontractors to report security breaches in sufficient time to allow the BAs to report back to the CEs. BAs may want to bind their subcontractors contractually to additional terms to help protect against security breaches such as requiring them to develop similar policies, procedures, and processes for investigating and reporting breaches.

BA Requirements for Compliance with HIPAA Security Rule

- BAs must be in compliance with the HIPAA Security Rule standards and implementation specifications for administrative, physical, and technical safeguards.
- Compliance means that BAs will likely need to do more in terms of securing PHI. Even though BAs are contractually required to implement appropriate "administrative, physical, and technical safeguards" to protect electronic PHI, the measures, policies, and procedures that a BA currently has in place may be insufficient for HIPAA compliance. The HIPAA Security Rule contains a series of very specific standards and implementation specifications. BAs must comply with each of the specific standards and implementation specifications under HIPAA.
- The first step in compliance is understanding the HIPAA Security Rule requirements and conducting a "gap analysis" to identify the areas where the BAs' information security systems and programs fall short of meeting the HIPAA Security Rule requirements. The Checklist preceding this chapter should help guide BAs in compliance efforts under the HIPAA security breach notification requirements.
- HHS annually issues additional guidance on technical safeguards that most appropriately implement the Security Rule. Therefore, compliance with the HIPAA Security Rule will require ongoing evaluation and updates.
- Consider HIPAA and HITECH compliance with respect to use of cloud-based resources offered by a cloud service provider (CSP). CSPs generally offer online access to shared computing resources with varying levels of functionality depending on the users' requirements, ranging from mere data storage to complete software solutions (e.g., an electronic medical record system), platforms to simplify the ability of application developers to create new products, and entire computing infrastructure for software programmers to deploy and test programs. Common cloud services are on-demand Internet access to computing (e.g., networks, servers, storage, applications) services. BAs that are CSPs or utilize CSPs in connection with their services for CEs need to take actions to ensure compliance with the Security Rule.
- BAs should ensure that contracts with subcontractors contain appropriate language to address information security and protect BAs from costs and liabilities associated with subcontractors' security breaches or other violations of contract terms related to information security. BAs should consider developing an information security due diligence questionnaire for potential subcontractors to evaluate their ability to protect PHI and other valuable data.

Statutory Liability for Business Associate Agreement Terms

- BAs are directly liable under HIPAA for violations of the terms of their BAAs.
- BAs should evaluate their current policies, procedures, and processes applicable to their ability to comply with each provision in their BAAs to ensure they are robust and will facilitate compliance.
- Training of personnel and evaluation of existing policies and procedures should be undertaken critically. Policies on employee sanctions for violations of HIPAA, the HITECH Act, and requirements in BAAs should be evaluated and strengthened.

BAA Compliance with HITECH Act Requirements

- BAAs should include the HITECH Act requirements for BAs, including compliance with the HIPAA Security Rule standards.
- With the public exposure that may result from breaches of unsecured PHI and the implications for their businesses, CEs typically require a broad range of business issues associated with the HIPAA security breach notification requirements. As a result, BAAs have become more complex. Responsibility for costs associated with security breaches and risk mitigation strategies in the event of a security breach are key issues in BAAs. CEs often press for broad indemnification from BAs. CEs may require BAs who are the subject of a security breach to make the required notifications on behalf of the CEs and to be responsible for all costs associated with a security breach.
- BAs should consider drafting their own form amendments and should create or revise their existing template BAAs to incorporate requirements of the HITECH Act. This will allow the BAs to create BAAs that contain the provisions required by law and yet are drafted to be more favorable and less burdensome to the BAs. This may help to minimize negotiation of terms that are not required by law, but that CEs will insert into form agreements to benefit the CEs and to reallocate risk to the BAs.
- These agreements should reflect the applicable obligations of the BAs. BAs should also consider inserting appropriate language to address information security and protect the BAs from costs and liabilities associated with subcontractors' security breaches or other violations of contract terms related to information security.

Other HIPAA Requirements

Pursuant to the HITECH Act, BAs must comply with a series of additional HIPAA requirements, including the HIPAA minimum necessary standards, the rules governing accounting of disclosures made from an electronic health record (EHR), prohibition on sale of PHI or an EHR, and conditions on marketing communications. These requirements are described below.

- BAs must comply with the HIPAA minimum necessary standards. These standards require BAs to limit the use and disclosure of PHI to the minimum quantity necessary to accomplish the purpose of the use or disclosure. Under the HITECH Act, as part of compliance with the minimum necessary requirements, BAs must make an initial determination whether the purpose for the use or disclosure could be practicably accomplished with a "limited data set." A limited data set is a set of data stripped of most identifiers. If BAs determine it is not practicable to use or disclose only a limited data set, then they must determine what constitutes the minimum necessary to accomplish the intended purposes of such use or disclosures. HHS has issued additional guidance on the minimum necessary requirements. Therefore, BAs' compliance with the HIPAA minimum necessary requirements will require ongoing evaluation and updates.
- CEs must provide an accounting to an individual of disclosures made from an EHR, including all disclosures for treatment, payment, or healthcare operations. Disclosures for such purposes have previously been exempted from the accounting requirements. If BAs maintain PHI for CEs in an electronic system that meets the definition of an EHR, then the BAs will be subject to these expanded accounting requirements. HHS will be issuing additional guidance on these requirements, and BAs will need to monitor this guidance to determine the ultimate impact on their operations.
- BAs cannot receive any direct or indirect remuneration in exchange for an EHR or PHI, with certain exceptions. BAs will have to evaluate their current arrangements involving transfer of PHI to determine if this prohibition affects them and if they have to revise their policies and procedures to reflect this requirement.
- BAs are subject to restrictions on communications about a product or service that encourage the recipient to purchase or use the product or service. BAs must evaluate their current arrangements involving such communications to determine if this prohibition affects them and if they need to revise their policies and procedures to reflect this requirement.

Steps for Compliance for Breach Notification

- Review existing policies and procedures to determine if they address security breach notification and identify modifications required for compliance with the regulations.
- Determine whether the organization is also subject to state breach notification requirements. If so, assess state law preemption and compliance issues as they relate to the HIPAA security breach notification requirements.
- Develop or refine security breach notification procedures to ensure that a centralized, coordinated security breach reporting system is in place. Consider the following:
- The procedures should designate the individual responsible for ensuring that breaches involving PHI are investigated fully. This individual also should have the ultimate decision-making authority for determining whether there has been a reportable breach. Procedures should provide for consultation with outside legal counsel as necessary for assistance with determining whether a reportable breach has occurred and, if so, when it was discovered.
  - The procedures should ensure that breaches are reported to the CEs without unreasonable delay. The procedures must ensure that individuals whose information has been compromised are appropriately identified, if possible, and reported to the CEs.
  - The procedures should require that employees be trained on the reporting procedures and the requirements for handling PHI.
  - The procedures should include appropriate sanctions for employees who mishandle PHI.
- Evaluate the PHI that the BA's organization controls and determine whether that PHI can be encrypted or destroyed so that it will be considered "secured" and not subject to the HIPAA security breach notification requirements.
- Maintain security breach notification policies as necessary to comply with HIPAA and state law. Obtain outside legal review of this policy to ensure that it complies with the HIPAA and state law requirements.
- Consider risk prevention and mitigation strategies for security breaches.
- Consider how the organization can enhance its security system to decrease the risk of breach of unsecured PHI.
- Evaluate the organization's insurance coverage to determine if it covers costs associated with security breaches of PHI.

Steps for Compliance with HIPAA Security Rule

- Conduct an organizational risk analysis to identify whether the HIPAA standards and implementation requirements are met. An audit tool should be used to assist with the initial risk assessment or gap analysis.
- The gap analysis should address each of the following implementation standards:
  - **Administrative safeguards**. HIPAA will require that BAs have certain administrative safeguards, including the following: (i) a security management process, (ii) an individual with assigned security responsibility, (iii) appropriate workforce security policies and procedures, (iv) policies and procedures for information access management, (v) a program of security awareness and training, (vi) security incident procedures, (vii) a contingency plan, and (viii) periodic evaluations of compliance with the HIPAA Security Rule.
  - **Physical safeguards**. HIPAA will require that BAs have certain physical safeguards, including implementation and maintenance of policies and procedures on facility access controls, policies, and procedures on workstation use, workstation security, and device and media controls.
  - **Technical safeguards**. HIPAA will require that BAs implement certain technical safeguards, including access controls, audit controls, integrity policies, person or entity authentication procedures, and transmission security procedures for PHI.
  - Develop written policies and procedures for each HIPAA standard listed above. *Please note that each of the implementation standards above has numerous requirements*. HIPAA requires that written policies and procedures be created that address each standard and each of the specific implementation specifications in the HIPAA Security Rule. These written policies and procedures are subject to record retention requirements of six years.
  - OCR has published guidance to assist CEs and BAs, including cloud services providers (CSPs), in understanding how they can use cloud computing technologies while complying with their HIPAA obligations. The guidance can be found here (https://www.hhs.gov/hipaa/for-professionals/special-topics/health-information-technology/cloud-computing/index.html).
  - Obtain review of policies and procedures to ensure legal compliance.
  - Train staff on HIPAA privacy and security rule requirements and the consequences of violation.

Additional BAA Terms

- Draft template BAAs that include all requirements under the HITECH Act. Require use of such templates by CEs where possible.
- Consider the complexities in negotiating BAAs with CEs, including the following:

 - CEs may conduct due diligence prior to contracting to determine whether BAs are HIPAA-compliant and whether a BA's security profile provides sufficient protection for PHI.
 - CEs may negotiate broad indemnification or cost-allocation provisions with their BAs to cover the CEs' exposure to costs associated with security breach notification requirements, potential reputational damage, and civil liability arising from BAs' breaches of unsecured PHI.
 - Check underlying services agreements for provisions addressing data privacy, security, and confidentiality to identify terms that conflict with the BAAs or place additional obligations on the BAs.
- Amend subcontractor agreements to address obligations that have been imposed on the BAs under its BAAs. Consider inserting appropriate language to address information security and to protect the BA from costs and liabilities associated with subcontractors' security breaches or other violations of contract terms related to information security. Seek assistance from legal counsel.

Considerations for Inventory HIPAA-Related Policies

- Periodically evaluate and update policies, procedures, and processes applicable to compliance with each provision in BAAs to ensure they are robust and will facilitate compliance.
- Include policies, procedures, and processes for other HITECH Act requirements as follows:
 - Accounting for disclosures made from an EHR
 - Minimum necessary disclosures/limited data set
 - Prohibition on sale of EHRs or PHI
 - Conditions on marketing communications
- Evaluate training procedures for personnel. Review and strengthen policies on employee sanctions for violations of HIPAA, the HITECH Act, or requirements in BAAs.

Summary

The preceding summary should guide companies by highlighting key provisions in HIPAA and the HITECH Act that apply to BAs with respect to achieving HIPAA compliance. BAs should also consult legal counsel to understand their obligations under HIPAA and the HITECH Act.

Chapter 17

Reducing Security Risks in Information Technology Contracts

CHECKLIST

Trade Secrets

- ☐ Stamp with "CONFIDENTIAL"
- ☐ Control physical access
- ☐ Use time stamps and ID logs
- ☐ Strong password requirements
- ☐ Encryption
- ☐ Firewalls
- ☐ Prohibited use of USB drives
- ☐ Isolate development and testing environments

Copyright

- ☐ Establish and communicate policy
- ☐ Mark with © symbol
- ☐ Mark with year of first publication
- ☐ Mark with name of legal owner
- ☐ Include textual marking in source code
- ☐ US copyright registration
- ☐ Register with US Customs

Joint IP

- ☐ "Clean room" protocols
- ☐ Isolate independent IP from joint IP

Embedded Open Source

- ☐ Policy against embedding open source
- ☐ Advance planning for correct embedding if at all

Internal Procedures

- ☐ Archive copies of each software version
- ☐ Verify company's right to use other IP
- ☐ Enforce security policies
- ☐ Appropriate use of computers
- ☐ Appropriate use of mobile devices
- ☐ Passwords

Policies After Infringement

- ☐ Audit rights
- ☐ "Phone-home" features
- ☐ Swift action upon infringement
- ☐ Terms for end of license
  - Uninstall program code
  - Destroy electronic copies
  - Return physical copies
- ☐ Insure against IP infringement

Employee Training

- ☐ Need to protect software
- ☐ How to protect software
- ☐ Responsibilities for protection during and after employment
- ☐ Exit interviews

Contractual Protections

- ☐ Proprietary information of former employer
- ☐ Assignment

- ☐ Prohibited use or disclosure of confidential information
- ☐ Noncompete agreements
- ☐ Nonsolicitation agreements

Nonemployees and Subcontractors

- ☐ Confidentiality agreements
- ☐ Need-to-know basis
- ☐ Work-for-hire agreements
- ☐ Assignment of all IP ownership rights

Software Distribution

- ☐ Only distribute object code, but if not:
  - Source code obfuscator
- ☐ Embed signature in code

License Agreements

- ☐ End-User License Agreement (EULA)
- ☐ Require acceptance of EULA
- ☐ Licensing in writing
- ☐ State clear terms and conditions
- ☐ No limited liability for misappropriation
- ☐ Breach results in breach of contract
- ☐ Breach results in IP infringement
- ☐ Specify narrow uses for IP
- ☐ No selling/transferring embedded software
- ☐ Prohibit reverse engineering
- ☐ Prohibit decompiling
- ☐ Prohibit discovering source code
- ☐ Prohibit discovering trade secrets
- ☐ Disclosure of accompanying documents
- ☐ Explicit statement of confidentiality

Nondisclosure Agreements (NDAs)

- ☐ Standard NDA for initial discussions
- ☐ After code delivery, license
- ☐ Perpetual trade secret confidentiality

Audit Rights

- ☐ Include audit rights
- ☐ Written certification by licensee officer
- ☐ Identify installations of software
- ☐ Retain certification copies for five years

Foreign Jurisdictions

- ☐ Distribute with care

Source Code Licenses

- ☐ Escrow the source code
- ☐ Limit release conditions
- ☐ Prohibit installation on network computer
- ☐ Licensee keeps copies in locked safe
- ☐ Prohibit copying onto removable media
- ☐ Limit personnel who can access code
- ☐ Third party: require written authorization
- ☐ No competitor access to code
- ☐ Keep logs of source code
- ☐ Use no open-source software
- ☐ Indemnify company from all infringement
- ☐ Warranties apply to unmodified software
- ☐ Prohibit IP rights in derivative works
- ☐ License to company for derivative works
- ☐ Total assignment of all IP is better
- ☐ Require specific security measures
- ☐ Right to audit licensee's use
- ☐ Strict confidentiality requirements
- ☐ Limited jurisdictions
- ☐ Limited remote access
- ☐ Risk of a "deemed export"

Best Practices and Guiding Principles

Effective intellectual property (IP) protection commences with a company's handling its own IP in a systematic and cautious manner. A proper foundation, both in educating employees and in maintaining best practices, is a necessary prerequisite for safe licensing of IP to prospective licensees. The licenses in particular require significant attention to detail in drafting the relevant portions to ensure that no unintended consequences result from loopholes or lack of clarity. This chapter outlines best practices—both within and outside a company—for controlling the handling and distribution of its IP.

Trade Secret Considerations

- All documents containing information that is not generally known to the company's competitors should be stamped "CONFIDENTIAL" or "TRADE SECRET." The primary means of protecting IP rights in software is through copyright and trade secrets. Trade secret protection can help ensure that the software, particularly source code, is always subject to rigorous confidentiality requirements.
- Where software or other design information may be readily observed, copied, or stolen, the company should control physical access to it. This includes time stamp and/or ID logs of those who have access to, and do access, the software.
- The company should adopt a strict system of data security measures, including strong password requirements, encryption, firewalls, and prohibited use of USB drives. The company should isolate the development and testing environments from the public Internet.

Copyright Considerations

- Establishing and communicating a policy for marking all copyrightable works, including design plans, source code, and software, should be a principal objective for the company. The company should mark all documents containing such information with the copyright symbol ©, year of first publication, and legal owner and include a textual marking of the same in the source code or other documents containing design information.
- Copyright protection arises as soon as the software is fixed in a tangible medium, which can be any medium that stores the software for an extended period of time, such as hard drives, flash drives, and CD/DVD ROMs. However, the software must be registered with the U.S. Copyright Office before the company can bring a claim of infringement, and companies may wish to implement a process for U.S. copyright registration of all versions of the software before any infringement occurs. Software should also be registered with U.S. Customs and Border Protection to prevent infringing copies from being imported.

Joint IP Considerations

In a joint development environment, the company should adopt "clean room" protocols and policies to ensure that existing and independently developed IP is isolated from new and jointly developed work product.

Policy on Embedded Open Source

The company should adopt policies to review the license terms of any open-source software components before employees and contractors may combine any open-source software components with the company's proprietary software. Failure to properly review the applicable open-source software's license terms could compromise IP rights in the proprietary code. There are "wrong" ways and "right" ways to implement open source in a proprietary environment, and the right way requires advance planning.

Internal Procedures

- The company should implement a system to archive copies of each version of the design information. For the purposes of documentation, this system establishes the overall course of development. Version control software automatically tracks and documents development.
- Verify and document the company's right to use the software and IP of others, including design information (including "cores"), graphics, artwork, software, and photographs.
- Implement and enforce company security policies to protect IP assets, including appropriate use of computer and mobile devices, and passwords.

Policies Following Infringement

- Through the use of audit rights and "phone-home" features, actively monitor the use of the company's IP by third parties and take swift action when infringement occurs.
- Ensure that the licensing agreements require the customer, on termination or expiration of the agreement, to uninstall the program code, destroy any electronic copies, and return physical copies of the code.
- The company should consider insuring IP against infringement.

Employees

Employee Training and Communication

- The company should train employees involved in developing, maintaining, and protecting its IP (including software, hardware designs, and any associated documentation) on the need to protect it, how to protect it, and their responsibilities in protecting it during and after employment.
- The company should take steps to secure its IP when employees depart the company by conducting exit interviews concerning IP issues, including discussion of inventions, and return of company property.

Contractual Protections

- The company should require that all employees and contractors execute appropriate confidentiality and proprietary rights agreements that limits the use and disclosure of confidential information. These agreements should also include required notification for certain protected disclosures under federal and state laws, including the U.S. Defend Trade Secrets Act of 2016 (DTSA). Failure to do so can limit the damages that the company can claim against employees who improperly disclose trade secrets.
- The company should require all new employees to acknowledge that they have not and will not use any proprietary information from any prior employer.
- The company should require all development personnel to execute such confidentiality and proprietary rights agreements that describe what inventions created by such personnel will be considered owned by the company. The requirements of this notice may vary from state to state, and the company may require different versions of these agreements for each jurisdiction where personnel perform their jobs. In addition, these agreements should: (i) explicitly state that work product that is copyrightable subject matter is "work made for hire" and that the employee waives any moral rights to the work product, and (ii) require a present assignment of all IP rights developed while they are in the employ of the company, and that such personnel will execute any necessary documents and/or allow the company to execute those documents on the employee's behalf.
- The company should ensure that the employee is required to return all materials containing company confidential information upon the termination of employment and that employee's duty of confidentiality continues after such termination.
- The company should also consider requiring certain employees to execute noncompete and nonsolicitation agreements.

Nonemployees and Subcontractors

- Subcontractors must be subject to appropriate confidentiality agreements. Nonemployees and subcontractors should have access only to modules as necessary to perform their tasks.
- All nonemployees and subcontractors, especially those engaged to create or contribute to any design information, should enter into work-for-hire agreements with an express assignment of *all* IP ownership rights to any deliverables created for the company and a perpetual license to all IP owned by the nonemployee or subcontractor created prior to the engagement with the company or outside the scope of the engagement.

Software Distribution

Object Code vs. Source Code

- It is important to distribute software only in object code form.
- If object code distribution is not possible, the company should consider:
  - Utilizing a source code obfuscator (i.e., scramble the symbols, code, and data of a software, rendering it impossible to reverse-engineer, while preserving the application's functionality).
- Embedding a "signature" that can be easily traced in the code (e.g., inserting a nonfunctioning block of code into functions or portions of the code that can later be used to verify whether portions of the code were copied, replacing a commonly used value (e.g., "0") with a symbol, number, or short text string).

Language for License Agreements

- An appropriate End-User License Agreement (EULA) should be included with the software and requires acceptance prior to installation or use of the software. Any licensing arrangements should be in writing and should set out the terms and conditions on which the IP may be used.
- In terms of limiting liability, ensure that misappropriation of IP by the customer is excluded from any damage cap or other limitation of liability clause. Breach of the license to the software should result not only in a breach of contract, but should also constitute an infringement of IP rights.
- License agreements should clearly and narrowly describe the specific uses the licensee can make of the software, including whether the software is subject to limitations such as specific hardware, locations, or servers on which it can be operated.

- For software embedded in hardware, the licensee should not be permitted to sell or otherwise transfer the hardware without the transferee's agreement to be bound by the license agreement.
- The license agreement should include express prohibitions against reverse engineering, decompiling, or otherwise acting to discover the source code and trade secrets of the design.
- Documentation accompanying the product frequently contains trade secret and other proprietary information. Its disclosure should be subject to a non-disclosure agreement (NDA) or other confidentiality obligations.
- In addition to a copyright notice, documentation should include a statement that the material is confidential, constitutes trade secrets of the licensor, and is provided solely in support of the licensee's use of the product.

Nondisclosure Agreements

- Initial discussions with potential licensees and provision of product documentation can be conducted under a standard NDA. Once any code is delivered, however, a license should be required.
- Confidentiality obligations with respect to trade secrets should be perpetual. NDAs and other confidentiality obligations frequently have time limits for their protections. While this may be appropriate for most confidential information, the presence of these limitations could result in waiver of trade secret protection. These provisions should be revised to ensure trade secrets will be protected as long as they are protected under applicable trade secret law.

Audit Rights

- The company should always try to include audit rights to ensure proper use of the software, although many companies will refuse to grant such rights or attempt to greatly limit the frequency and scope of them. Use a third-party auditor who specializes in conducting compliance audits and determines its fees solely as a function of instances it uncovers in which the licensee has used the company's product in violation of the license agreement (e.g., the auditor receives a percentage of revenue generated by the excess use, but no other compensation).
- In addition to audit rights, the company should require that, on a periodic basis, an officer of the licensee certifies in writing that all use of the company's product is in compliance with the terms of the agreement. In particular, the certification should identify all installations and uses of the software. Copies of the certifications should be retained until at least five years after expiration of the license agreement.

Foreign Jurisdictions

- Distribution in foreign jurisdictions should be done with care to ensure the relevant locations respect IP rights.
- Some technologies included in products, including those in software, hardware, or firmware, may be subject to certain restrictions and requirements under the export control laws of the United States and possible import restrictions in other countries. The company should always include an obligation that the licensee will comply with all such requirements and restrictions. Be aware that access to such technology in the United States by foreign nationals may constitute a "deemed export."

Source Code Licenses

Escrow the Source Code

- Providing licensees with access to source code is strongly discouraged. If a licensee insists on access, the initial response should be to, at most, escrow the source code with an approved escrow agent. The release conditions from the escrow should be limited to voluntary bankruptcy, the company's decision to cease support of the entire product line, and other appropriate narrow conditions.

Language for Source Code License Agreements

- If source code is to be licensed, it must be done under a specifically drafted source code license agreement that, among other things, does the following:
  - Prohibits the licensee from installing the source code on any networked computer (whether an internal or external network).
  - Requires the licensee to keep physical copies of the source code in a locked safe when not in use.
  - Prohibits copying the source code onto any form of removable media (e.g., USB fobs, CDs, DVDs, removable drives).
  - Strictly limits the licensee personnel who can access the source code.
  - Prohibits access to the source code by any third-party contractor without the company's express written authorization. At minimum, competitors should be precluded from ever accessing the source code.
  - Requires retention of complete and accurate logs of all access to and use of the source code.
  - Strictly precludes the licensee from using any open-source software in connection with the source code.

- Requires the licensee to indemnify the company from any and all infringement claims that may arise from their revisions to the source code.
- Makes clear that any warranties, indemnities, and support obligations are applicable only to the unmodified version of the software. Once a licensee modifies the source code, the obligation will no longer apply.
- Prevents the licensee from applying for or obtaining any IP rights in any derivative works.
- Includes express contractual provisions preventing the licensee from ever enforcing any rights it may have in the derivative works against the company or its customers.
- Includes a broad, irrevocable license from the licensee to the company for all derivative works. An outright assignment of IP rights would be preferred.
- Requires licensee to follow specific information security measures in handling and using the source code.
- Includes the company's right to audit the licensee's use of the source code, including the use of third-party auditors.
- Clearly and narrowly defines the licensee's uses of the source code.
- States that all licensee personnel coming in contact with the source code must be bound by strict confidentiality agreements.
- States that licensees should be strictly limited in the jurisdictions in which the source code may be used. As noted above, some jurisdictions do not respect or protect IP. In addition to physical transfer of the software to other jurisdictions, the license should also limit remote access to the software in those jurisdictions (e.g., the software is located in the United States, but accessed in Russia).

Summary

The dynamic environment of technology development warrants genuine caution on behalf of companies participating in this space. In many cases, a company's IP can be among its most valuable assets. Anticipating the potential risks associated with IP can prevent a host of issues from arising after the opportunity to protect IP has passed.

Chapter 18

Website Assessment Audits

CHECKLIST

Site Evaluation

- ☐ Target audience?
- ☐ Accessibility to all users?
- ☐ Users in the United States or abroad?
- ☐ Process transactions?
- ☐ Products or services sold?
- ☐ Forms?
- ☐ Enter data or access database?

Domain Names

- ☐ Proper registration
- ☐ Proper entity listed as owner
- ☐ Trademark due diligence
- ☐ Search for "cybersquatters"
- ☐ Using domain brokers

Use of Third-party Trademarks

- ☐ Written permission prior to use
- ☐ Written permission for quotations

- ☐ Meta tags
- ☐ White text on white background
- ☐ Microscopic type

Hyperlinks

- ☐ Trademark or logo symbols
- ☐ Interstitial notice in terms and conditions
- ☐ Written linking agreement
- ☐ No implicit endorsements
- ☐ No representations about linked sites
- ☐ No framing without permission
- ☐ Written permission for deep linking

Content

- ☐ Development agreement
- ☐ Agreements with independent contractors
- ☐ Employment agreements
- ☐ Other agreements regarding the site
- ☐ Website has right to use content
- ☐ Third-party content providers
- ☐ Photographs

Visitor Uploads

- ☐ Submission agreement
- ☐ Visitor accepts liability associated with upload
- ☐ Chat/discussion room disclaimers
- ☐ DMCA requirements
  - Permit operator to terminate service
  - Do not interfere with protection of IP
  - Agent to notify if infringement
- ☐ File agent name at Copyright Office

Internet Laws

- ☐ Spamming
- ☐ Sales
- ☐ Advertising
- ☐ COPPA

Terms and Conditions

- ☐ Accessible from home page
- ☐ Accessible by link
- ☐ Methods to determine visitor assent
  - Required online registration
  - Required acceptance
  - Prominent notice
  - Basic notice
- ☐ Changes to legal notices
- ☐ Applicable law and venue
- ☐ Arbitration clause

Data Security and Privacy

- ☐ Privacy policy?
- ☐ Accessible from home page
- ☐ Links to terms and conditions
- ☐ Employees follow policy
- ☐ Third-party online privacy certification
- ☐ Agreement with hosting provider
- ☐ Firewall

Insurance

- ☐ Intellectual property infringement
- ☐ Invasion of privacy
- ☐ Defamation
- ☐ Personally identifiable information
- ☐ Protected health information
- ☐ Personal financial information
- ☐ Misuse of information by site
- ☐ Misuse of information by employee

Additional Concerns

- ☐ Record of modifications to T&C
- ☐ Copyright notice on site

Overview

Launching and maintaining a website require substantial background knowledge on the associated risks and limitations. Issues range from evaluating the integrity of prospective domain names and ensuring the integrity of the company's domain name; using third-party trademarks and linking to their protected sites; and navigating through the vast network of federal and state regulations governing the practice of website development. This chapter discusses some of the main issues to consider in launching and maintaining a website.

While it may seem that many website are relatively straightforward, the reality is that most websites are highly complex combinations of programming and content from dozens of third-party sources. In addition, the website is usually being hosted by yet another third-party vendor. There may also be agreements with a further range of vendors to provide various services in support of the website. Those agreements can range from payment processing contracts to order fulfillment to protection from denial of service attacks.

Key Issues and Guiding Principles

Evaluate Your Website

- Who is the website's targeted audience? This audience could be vendors, dealers, resellers, strategic partners, consumers, children, regulators, or search engines.
- Is the entire website accessible to all users, or is access to certain portions limited to classes of users with varying user rights? Additionally, are the website's users generally located inside or outside the United States? If outside, can the locations be identified where users outside of the United States (generally) access the site? If so, what percentage of all site users are non-U.S. users?
- Does the website process any transactions? These may include credit applications, sale of goods, and user registrations.
- Are any products or services sold or made available on the website? If so, they should be identified. Products or services may include tangible products, insurance, securities, information, and financial services.
- Does the website use any forms? If so, how does it use forms? Uses of forms could include product ordering, collection of personal information, etc.
- Does the website enter data into or access a database? If so, does the website owner have the proper licenses to use such database, including tables, table definitions, entry forms, report engines, etc.? Know the location of the database and compile a list of who has access to the database if it is shared.

Domain Names

- **Proper registration (whois.net).** The domain owner should confirm that the registration information is accurate and should promptly make any changes if any of the information is not accurate at present or in the future. Additionally, the owner should confirm that the proper entity is listed as the owner of the domain name.
- **Trademark due diligence (USPTO.com).** Actions for trademark infringement are the primary source of litigation involving websites. As such, any use of a third-party's marks should be carefully reviewed. If any related marks are found, it may be advisable to consult with trademark counsel to determine the impact of these results.
- **Registration of likely related names.** A full U.S. trademark search will provide names similar to the registered domain name and may reveal that some domain names were registered in bad faith by potential "cybersquatters." The trademark owner may bring suit for trademark infringement for damages to compel a cybersquatter to transfer the domain name. The trademark owner may also seek resolution under the Uniform Dispute Resolution Policy ("UDRP") of the Internet Corporation for Assigned Names and Numbers ("ICANN") (www.icann.org).
- If a domain name was registered in good faith by a third party, one may use the services of a domain broker (e.g., greatdomains.com) to negotiate and handle the details of acquiring the domain name.

Use of Third-Party Trademarks

- The domain owner should obtain written permission from all parties prior to using their name or logo on its website. Written consent should also be obtained for any quotations from clients or other third parties, as well as the use of their names.
- Many courts have held that the use of meta tags, white text on white background, and microscopic type of a competitor's trademarks constitutes infringement. A meta tag is a special text that is inserted into a HTML document or web page and is not visible when the web page is displayed (unless the source code is viewed). Meta tags contain information relating to the website including keywords that describe the page's content. To drive traffic to their sites, website owners and developers may use competitors' trademarks in their meta tags.

Hyperlinks

- If the third party's trademark or logo is used as a hyperlink and it has a symbol used to indicate the mark is registered (®) or that trademark rights are claimed in it (™), those symbols should be used. The website terms and conditions should include an interstitial notice (a statement that all trademarks or service marks are the property of their respective owners).

- A written linking agreement should appear on any site, setting out the parties' respective rights and obligations.
- Do not use another business' name, trademarks, or logos improperly to imply the endorsement of that business. Where no endorsement is intended, include language in the website terms and conditions stating that neither the operator of the website nor the linked business intends any endorsement of the other because of the link.
- **Disclaimer of liability and control**. A website's terms and conditions should make clear that the operator has no control over the third party's site and disclaim all liability for a visitor linking to the third party's site via the hyperlink provided.
- Include language in the website terms and conditions to the effect that the operator of the website makes no representations about linked sites or their owners, products, or services. Conversely, review the terms and conditions for each hyperlinked website to determine if it has any limitations or prohibitions on hyperlinking.
- **Framing**. A web designer can create multiple, scrollable windows or frames that appear in the user's browser. Designers can thereby use frames to incorporate entire third-party websites, or only portions of them, into their sites and surround them with their own advertising and promotions. ***Never*** engage in framing without the express permission of the operator of the framed site.
- **Deep linking**. Written consent from a website owner should be obtained if the website wants to deep-link to another site. Consider including in the terms and conditions a restriction on allowing other sites to deep-link to the website.

Content

- The domain owner must be the owner or licensee to each piece of content incorporated into the website. A careful review of the following documents should be made for their intellectual property or licensee provisions:
  - The development agreement.
  - Agreements with independent contractors who contributed to the development of the website.
  - Employment agreements of employees who contributed to the website. Issues frequently arise when employees contribute content to a website outside the normal scope of their employment. This work is not "made for hire" under the Copyright Act and, therefore, is not owned by the employer.
  - Any other agreements related to the website.

- Ensure through written agreements with all third-party content providers that:
  - The website has the right to use the content as intended.
  - The third-party content providers have the right to convey such rights to the website owner without violating the copyrights or other rights of third parties who may have rights in or to the content (e.g., other authors, developers).
- Photographs may violate an individual's privacy and publicity rights. Each employee pictured should be required to execute a model release for the use of his or her image on the website.

Visitor Uploads

- All visitors who upload material should accept submission agreements or accept such terms in the terms and conditions whereby the visitor accepts all liability for any infringement, misuse, claims, or issues otherwise associated with such upload.
- **Chat/discussion rooms disclaimers**. The terms and conditions should prohibit unlawful, offensive, harassing, discriminatory, or other improper conduct. The terms and conditions should also give the website owner broad rights with respect to enforcing these prohibitions and restrictions.
- **The Digital Millennium Copyright Act ("DMCA")**. Title II of the DMCA is entitled "Online Copyright Infringement Liability Limitation." In brief, Title II provides virtually complete protection from damages resulting from copyright infringing material placed on or passed through a website by its users. A website must include:
  - Terms and conditions of use for its users that permit the operator to terminate service if a user commits copyright infringement.
  - Terms and conditions that do not interfere with any technical measures taken by the copyright owner to identify and/or protect its works.
  - Identification of a designated agent to receive notification of potential copyright infringement. The designated agent should be identified on the appropriate filing with the U.S. Copyright Office.

Applicable Internet-Specific Laws

- **Spamming**. If an e-mail address is used for the registration process and the website owner intends to use the e-mail addresses collected for any purpose other than sending the username and password to the visitor, it may run afoul of one or more state anti-spam laws. If additional uses are anticipated for the e-mail addresses, the visitor should be advised of such uses and afforded the opportunity to opt out of the mailings. Avoid sending unsolicited e-mail to visitors without their permission.

- **Sales.** Does the website sell products and collect taxes in states with a nexus to the site operator, such as states with a physical presence?
- **Advertising.** Is the website compliant with the Federal Trade Commission's rules regarding bait and switch, catalogs, children's advertising, comparative ads, contests and sweepstakes, credit, disclosures and disclaimers, endorsements and testimonials, food ads, franchises and business opportunities, free claims, guarantees, Internet advertising, leasing, mail order advertising, pricing, and rainchecks?
- **COPPA.** The Child Online Privacy Protection Act (COPPA) applies to commercial websites or online services (i) directed to children under the age of 13 that collect personal information from children or (ii) that operate general audience websites and have actual knowledge that they collect personal information from children, including hobbies, interests, and information collected through the use of cookies or other types of online tracking mechanisms—whenever such information is associated with a particular child. If COPPA applies to a site, it spells out what the operator of the site must include in a privacy policy, when and how to seek verifiable consent from a parent, and what responsibilities the operator has to protect children's privacy and safety online. Complying with the requirements of COPPA can be costly and time-consuming. The Federal Trade Commission (FTC) has created an excellent guide for website operators to assist them in complying with the requirements of COPPA: *How to Comply with the Children's Online Privacy Protection Rule.* It is available on the FTC's website at www.ftc.gov/bcp/conline/pubs/buspubs/coppa.htm.

Terms and Conditions

- Terms and conditions should be accessible from the home page and accessible via a "Terms and Conditions" link. Generally, the terms and conditions will also link with the privacy policy.
- Enforceability of the terms and conditions generally depends upon whether the visitor to a site has seen the terms and conditions. Generally, there are four methods to indicate a visitor's assent to the terms and conditions:
  - **Required online registration.** Make sure visitors go through a registration process where they are required to view the terms and conditions before clicking on an "I accept" or "I reject" box to indicate that they have reviewed the terms and conditions.
  - **Prominent notice and required acceptance.** When a visitor first lands on the page, in order to access the site, a link to the terms and conditions is provided and the visitor is asked to click on an "I accept" or "I reject" link to access the content.
  - **Prominent notice.** The first page of the site has prominent language regarding the terms and conditions (with a hyperlink to the actual contract language), and acceptance is implied through conduct. The relevant

language could be: "USE OF THIS WEBSITE IS SUBJECT TO CERTAIN *TERMS AND CONDITIONS*. YOUR CONTINUED USE OF THIS SITE INDICATES YOUR ACCEPTANCE OF THOSE TERMS AND CONDITIONS."

  - **Basic notice.** A hyperlink to the terms and conditions is included at the bottom of the first page of the website: "Click here for Terms and Conditions." However, some courts have found this type of basic notice insufficient to form a binding agreement with website visitors. Although this method is very popular and found on many of the most frequently visited websites, it is the least likely to be found enforceable, because there is no evidence that the visitor agreed to the terms and conditions or even read them.

- Every time there is a change to the legal notices, the date of such change should be recorded and copies of prior posted versions should be retained for at least six years.
- The terms and conditions should specify the applicable law and venue for disputes related to the website.
- The terms and conditions should contain an arbitration clause requiring a visitor to seek binding arbitration to settle disputes.

Data Security and Privacy

- Is there a privacy policy? Some states, for example, California, now require websites to have a posted privacy policy.
- The privacy policy should be accessible from the homepage and accessible via a "privacy policy" link. Generally, the privacy policy will also link with the terms and conditions.
- Is the privacy policy strictly followed and do the employees know and understand it?
- Is it appropriate to obtain third-party online privacy certification?
- Is there an agreement with the website hosting provider regarding the use and protection of the website's visitor's information?
- Does the website use appropriate firewall technology?

Insurance

- Does the website owner have the proper insurance based upon the online activities of the website? Consider insurance coverage for:
  - intellectual property infringement
  - invasion of privacy
  - defamation
  - breach of personally identifiable information, protected health information, or personal financial information
- Misuse of information by site or employee

General Considerations

- Keep accurate records of modifications and amendments to the terms and conditions and privacy policy.
- Although use of a copyright notice is no longer required under U.S. law, its use provides the copyright owner with a number of benefits (e.g., avoiding claims of innocent infringement). The copyright notice should be placed on the website such that it gives reasonable notice of the claim of copyright. This notice should be visible to an ordinary user of the site under normal conditions of use and should not be concealed from view upon reasonable examination.
- Websites, even simple ones, are a complex amalgamation of content and programming. One of the key questions every business should ask is "can I take this website elsewhere?" The problem is that the website may have been developed by one vendor, hosted by another, and contain third-party content and programming from dozens of other licensors and vendors. If the business decides to move the website to another host or, potentially, host the website on its own internal servers, does the business have all rights necessary to do that? The answer can be incredibly complex. Every business should ask that question now and do an audit of its website to determine whether it has all necessary rights or not.

Summary

Developing a website requires significant consideration of the background risks and potential liabilities associated with participating in the Internet community. By asking the questions outlined in this chapter, a company is in the best position to evaluate its own site's potential for development, inherent risks, and potential conflicts with regulations and governance for the Internet.

Chapter 19

Critical Considerations for Protecting IP in a Software Development Environment

CHECKLIST

Key Issues

- ☐ Definitions
- ☐ Risk of contracting

Vendor Due Diligence

- ☐ Put vendors on notice
- ☐ Security standards
  - Gramm–Leach–Bliley
  - HIPAA Security Rule/HITECH Act
  - FFIEC Guidance
  - States
  - Federal Trade Commission
- ☐ Diligence should cover:
  - Criminal convictions
  - Litigation

    - Regulatory and enforcement
    - Breaches of security
    - Breaches of health information
    - Adverse audits
    - Use of parties outside the United States
- ☐ Standardized questionnaire
    - Corporate responsibility
    - Insurance coverage
    - Financial condition
    - Personnel practices
    - Information security policies
    - Physical security
    - Logical security
    - Disaster recovery
- ☐ Business continuity

Treatment of Data

- ☐ Maintain data as confidential
- ☐ Liability for unauthorized disclosures
- ☐ No data removed by vendor

Administrative Security

- ☐ Written privacy policy
- ☐ NDAs for personnel with access
- ☐ Trade secrets
- ☐ Written security plan
- ☐ Encryption
- ☐ Procedures for removable media
- ☐ Permission settings and restrictions
- ☐ Separate networks with respect to access
- ☐ Permanent logs of any access
- ☐ No unauthorized access to client data
- ☐ No installation or removal of programs
- ☐ Require reasonable security
- ☐ Vendors abide by regulatory framework
- ☐ Document access by vendors

Technical Security

- ☐ Enable use of firewalls
- ☐ Ensure secure Internet access

- ☐ Consider disconnecting computers
- ☐ Encryption
- ☐ Procedures for data in transit
- ☐ Separate testing from production

Personnel Security

- ☐ All aware of security requirements
- ☐ Client can request removal of personnel
- ☐ Prescreening
- ☐ Control over access
- ☐ Review of materials taken outside

Subcontractors

- ☐ Identified in writing
- ☐ Client right to approve/reject
- ☐ Vendor accepts liability
- ☐ Mirror PSA

Scan for Threats

- ☐ Prohibit install
- ☐ Accessible by link
- ☐ Methods to determine visitor assent
  - Required online registration
  - Required acceptance
  - Prominent notice
  - Basic notice
- ☐ Changes to legal notices
- ☐ Applicable law and venue
- ☐ Arbitration clause

Data Security and Privacy

- ☐ Privacy policy?
- ☐ Accessible from home page
- ☐ Links to terms and conditions
- ☐ Employees follow policy
- ☐ Third-party online privacy certification
- ☐ Agreement with hosting provider
- ☐ Firewall

Insurance

- ☐ Intellectual property infringement
- ☐ Invasion of privacy
- ☐ Defamation
- ☐ Personally identifiable information
- ☐ Protected health information
- ☐ Personal financial information
- ☐ Misuse of information by site
- ☐ Misuse of information by employee

Additional Concerns

- ☐ Record of modifications to T&C
- ☐ Copyright notice on site

Overview

Businesses must be rigorous in entering into vendor relationships in which sensitive information will be placed at risk. Security requires a unified approach, including but not limited to security policies, employee education, use of security technology, performing security audits, and addressing security in contracts with business partners and other vendors. Information security can be divided into three categories—administrative, technical, and physical. In this chapter, we evaluate tools that businesses can immediately put to use to substantially reduce the information security threats posed by their vendors and business partners, to ensure proper diligence is conducted and documented, and to provide remedies in the event of compromised security.

Key Issues and Guiding Principles

- **Definitions.** The definition of *data* should include all information to which the vendor may have access, including the company's customer information, the company's proprietary and confidential information, and any other nonpublic information provided by the client to the vendor, including its intellectual property and business information. In many instances, a company's proprietary and confidential information is *the* most important asset of the company.
- **Assess the risk of contracting.** Does the risk of involving another party outweigh the benefits of services provided by that third party? If not, an agreement should be in place any time the third party will have access to data.

Vendor Due Diligence

- Companies should put all vendors on notice that their security policies and procedures will in part determine whether any particular vendor shall be selected to have access to data.
- A company must also consider all applicable security standards, including:
  - Gramm–Leach–Bliley Act (a federal law directed at the protection of nonpublic, personally identifiable financial information)
  - HIPAA Security Rule/HITECH Act (discussed in Chapter 16)
  - Federal Financial Institutions Examinations Council (the FFIEC) Outsourcing Technology Services Guidance
  - States (e.g., California, Massachusetts, Nevada)
  - Federal Trade Commission
- Diligence should cover the following topics:
  - Criminal convictions
  - Litigation
  - Regulatory enforcement actions
  - Breaches of security or health information
  - Adverse audits
  - Use of affiliates, subsidiaries, contractors, and vendors outside the United States
- Use a standardized questionnaire with the vendor covering the following topics, rather than relying on an ad hoc process to evaluate the integrity of vendor security:
  - **Corporate responsibility**. Are there any criminal convictions, recent material litigation, or instances in which the vendor has had a substantial compromise of security, privacy violations, adverse audit results?
  - **Insurance coverage**. What types of coverage does the vendor have? What are the coverage limits and other terms? Is the coverage based on claims made or occurrences? In particular, does the vendor have cyber liability coverage?
  - **Financial condition**. Is the vendor a private or public company? Review copies of the most recent financial statements.
  - **Personnel practices.**
  - **Information security policies.**
  - **Physical security.**
  - **Logical security.**
  - **Disaster recovery and business continuity.** What are the vendor's business continuity plans and disaster recovery plans? When was its last test? When was its last audit? Were there adverse findings in the audit? Have deficiencies been corrected?

Treatment of Data

- The vendor must maintain all information it receives from the client as confidential. The vendor should be liable for any unauthorized disclosures or use of data by its personnel or a subcontractor. No data should be removed by a vendor.

Physical Security

- Consider securing the physical grounds where data is kept, including using locks and restricting access. Servers should be separated from highly sensitive information in rooms with added security and restricted access.
- Consider requiring security guards and cameras.
- If a vendor is responsible for the storage and protection of a company's data, the company should ensure that the vendor stores the information in a physically and logically secure environment that protects it from unauthorized access, modification, theft, misuse, and destruction.

Administrative Security

- A written privacy policy should be in place to require the vendor—including all personnel and subcontractors—to comply with the requirements and obligations of the policy.
- All personnel who have access to data should be required to execute a nondisclosure agreement.
- Similar protections should be used with respect to trade secrets.
- A written security plan should incorporate the use of passwords and portable computing devices (e.g., laptops and smartphones), as well as portable storage devices (e.g., USB drives). The plan should also mandate the use of encryption on any device that provides or enables access to any confidential or sensitive information, including any client intellectual property. In the midst of developing a plan, one should be constantly aware of the ability to remotely track or wipe portable devices when they are reported lost or stolen. Consider procedures for and management of the use of removable media, including keeping logs of use.
- Use permission settings and restrictions. Users and vendors should not have rights or access to any information, systems, or programs that they should not have.
- Consider separating networks or systems to which vendors do not have access from systems to which they do not require access.
- Keep permanent logs of any access to the information, including logs of all activity and system, program, or user faults.

- Vendors shall not allow unauthorized access to companies' data. Companies should not allow their users or vendors to install or remove any programs without their authorization. Require vendors that hold or access data to implement reasonable security procedures.
- Require vendors, at a bare minimum, to follow and abide by all regulatory requirements, including:
  - Gramm–Leach–Bliley
  - HIPAA Security Rule/HITECH Act
  - FFIEC Guidance
  - States (e.g., California, Massachusetts, Nevada)
  - Federal Trade Commission
- Document all equipment and systems to which vendor has access and regularly ensure that it is protected and not tampered with.

Technical Security

- Enable the use of firewalls to protect computers and networks.
  - Require that when vendors access company's systems and networks that they use firewalls, antivirus programs, and VPNs.
- Ensure that all connections to the Internet are secure. In the alternative, if an Internet connection is not needed for certain computers, consider disconnecting them.
- Implement intrusion detection systems.
  - Encrypt all sensitive data.
  - Clients should have procedures regarding data in transit, including:
  - Encryption of all information in transit.
  - Protection by vendor from unauthorized access, misuse, and disclosure of data in transit.
  - Logging of all data in transit.
- Maintain separate environments for testing and production.

Personnel Security

- Ensure that all vendor personnel are aware of the security requirements, including access and disclosure restrictions. Consider regular or annual trainings and reminders.
- Companies should have the ability to request the removal of any personnel that present a security threat.
- All vendor personnel should be screened prior to gaining access to any data. Screening should include character references, confirmation of claimed academic and professional qualifications, and an identity check.

- **Control over access.** Company personnel should be required to escort vendors when moving throughout companies facilities. All contractor personnel should have identification that indicates their status, security clearance, and access.
- Review and scan any materials or equipment that vendor personnel take out of companies' facilities.

Subcontractors

- Proposed subcontractors should be identified in writing to companies prior to being staffed on an engagement. The company should reserve the right to approve or reject any proposed subcontractors.
- Vendors should accept responsibility for any subcontractor liability. Liability and confidentiality provisions in the subcontracting agreement should mirror the liability provisions in the professional services agreement.

Scan for Threats

- Companies should have a provision in agreements with vendors that prohibits vendors from installing any program or code that would enable vendors to access company systems or that would otherwise impair or harm the system.
- Companies should require that vendors scan any electronic file prior to delivery to companies for any viruses or any program or code designed to disrupt, harm, or otherwise impede company systems.
- Monitor systems regularly for compliance and any threats. Routinely test systems and facilities for vulnerabilities.

Backup and Disaster Recovery

- All information should be regularly backed up and securely stored in a remote location. The backup should use strong encryption to protect data. Know where the backup site is and how secure it is. Beware of vendors that use offshore backup facilities, particularly in jurisdictions with limited or nonextant intellectual property, privacy and security laws and regulations. It may be impossible in those jurisdictions to be certain that residual data is actually and completely deleted from local storage. In some cases, local laws may require retention of company data for years after expiration or termination of the vendor agreement.
- Ensure the data is only transmitted via secure methods.

Confidentiality

- Confidentiality provisions should incorporate all types of information to which companies may potentially provide vendors access.

Security Audits

- During the term of the agreement, clients should have the right to conduct, or have conducted on their behalf, an audit of vendors' security measures.

Warranties

Vendors should warrant that:

- They have the authority to enter into this agreement, to perform all the services it requires, and to provide all necessary deliverables.
- They will follow and abide by or, if applicable, maintain and enforce any physical security procedures with respect to the access and maintenance of the client's data. Vendors should comply with the best industry security practices.
- They will not make any deceptive claims regarding the privacy or security they provide regarding the data.
- As applicable, their compliance with and that they shall abide by company's privacy policy and all federal, state, and local laws, rules, and regulations, including:
  - Gramm–Leach–Bliley
  - HIPAA Security Rule/HITECH Act
  - FFIEC Guidance
  - States (e.g., California, Massachusetts, Nevada)
  - Identity theft regulations
  - Federal Trade Commission Red Flags Rule (a federal law designed to protect against identity theft)
  - Vendors will not use, transmit, or make available client data outside of the United States without the client's prior written authorization.
- Consider requiring a warranty to reduce risk in vendor remote facilities. Language like the following is typically used:

  Vendor represents and warrants that all vendor systems used in providing the services that access customer systems and the data therein shall be configured as follows: (i) all print screen, screen capture, and similar functionality shall be disabled; (ii) no customer data will be cached on vendor systems or transferred to any form of local storage media; (iii) the vendor systems shall not be capable of printing any customer data or other confidential information; (iv) all USB, FireWire, and other similar ports shall be disabled; and (v) all wireless services (e.g., WiFi, Bluetooth®) shall be disabled. In addition, vendor shall not permit any recording devices (e.g., cameras, smartphones, audio records, video recorders) of any kind in the areas of vendor's facilities where the systems are located that access and display customer data.
- Their warranties are not confined to the warranties section of the agreement.
- Their responses to the due diligence must be true and accurate.

- **Indemnification**. Vendors shall indemnify clients from any third-party claims related to the vendors' breach of confidentiality or its failure to comply with security requirements. That is, the vendor should protect the business from lawsuits and other claims that result from the vendor's failure to adequately secure its systems.

Limitation of Liability

- Typically, agreements limit the vendor liability to the amount of fees paid by the client, the total value of the agreement or a particular work order, or to a predetermined amount. There should be carve-outs for:
  - Indemnification of the intellectual property.
  - Breach of confidentiality.
  - Use of client's name.
  - Misappropriation of intellectual property.

Termination

- The company should be able to immediately terminate the agreement due to the vendor's breach or for compliance failures.

Security Breach Notification

- The agreement should specify who is responsible upon a breach of security regarding control of the notice to affected parties and the costs.

Insurance

- Ensure that vendors have adequate insurance in case of a security breach.

Destruction of Data

- Ensure that the vendor destroys all data and certifies its destruction, including ensuring that all hardware has been sanitized or wiped clean. Procedures should be set forth for the secure disposal of media no longer needed.

Additional Considerations

- Include language in the agreement that vendors shall take all reasonable measures to:
  - Secure and defend their systems and facilities from unauthorized access or intrusion.

  - Periodically test their systems and facilities for vulnerabilities.
  - Immediately report any breaches or potential breaches of security.
- Cooperate with the client on any regulatory audits or in reviewing information on security policies and procedures.

Summary

Information security protections as described in this chapter are important because they protect valuable assets in business. They establish due diligence, protect business reputation, and help companies avoid public embarrassment while minimizing potential liability. Using a three-step approach, including establishing vendor due diligence, contractual protections, and procedures for handling information, companies are able to secure their information.

Chapter 20

Transactions Involving Financial Services Companies as the Customer

CHECKLIST

Form and Type of Agreement

- ☐ Company's form or vendor's form

Definitions

- ☐ Definition of "confidential information"
  - Personally identifiable information
  - Trading and account information
  - "Insider information"
- ☐ Definition of "aggregated data," if applicable

General Requirements

- ☐ Include all standard requirements discussed elsewhere in this book for the particular type of contract under consideration

- ☐ Strong confidentiality clause
  - Perpetual protection for personal information
  - Ongoing protection for trade secrets
- ☐ Compliance with customer's privacy policy, including updates
- ☐ Control over notices for data breaches
  - Reimbursement for costs, notice, investigation, identity theft insurance
- ☐ Avoid data aggregation rights
  - Obligation to cleanse/scrub the data
  - As-is
  - Indemnity for failure to cleanse and all use of data
- ☐ Information security
  - Best industry practices
  - Compliance with applicable laws and regulations
  - Prompt reporting of potential or actual breaches
  - Maintain and provide log files and other forensic evidence
  - Audit rights
  - Testing, including penetration testing
  - Right to SSAE or similar audits
  - Requirements for secure deletion and data removal
- ☐ Background checks
- ☐ Indemnification for breach of confidentiality
- ☐ Breaches of confidentiality and indemnifications obligations excluded from limitations of liability
- ☐ Audit rights
  - Security
  - Contract performance
  - Confirm charges and fees
  - Regulators
- ☐ Termination for regulatory issues
- ☐ Reject vendor audit rights in favor of off-site record review
- ☐ Review pricing and tying arrangements between and among products and services
- ☐ Compliance of software and services with relevant laws and regulations
  - Right to updates without charge
- ☐ Limit subcontractors
  - Offshore
  - Due diligence
  - Potential separate NDA

Techniques

- ☐ Be ready to explain the unique legal and regulatory requirements.

- ☐ Be familiar with the *Federal Financial Institutions Examination Council Handbook*. This should be your first stop in understanding your obligations.
- ☐ Review checklist of regulatory considerations at the end of this chapter. Update the checklist for your company's specific regulatory requirements.
- ☐ Make your own checklist of key issues.

Overview

This chapter discusses the unique challenges faced by financial services companies (e.g., banks, broker-dealers, insurance companies) when they enter into technology contracts. As with any customer entering into a contract for the use or acquisition of technology, financial services companies must be concerned with warranties, indemnities, scope of license, statements of work, intellectual property ownership, and the dozens of other issues common to agreements of this kind. These issues are discussed in depth in other chapters of this book. Here, however, we are going to focus on the additional, unique risks and concerns financial services companies must address in contracting for technology.

There are essentially two driving forces behind these unique risks and concerns. First, these types of entities are in the business of handling very sensitive data. As such, their attention to issues such as confidentiality and information security is greatly heightened. Second, financial services companies are some of the most highly regulated entities in the world. They are subject to a wide range of state, federal, and, in some instances, international laws and regulations regarding almost every aspect of their operations. In addition, they receive frequent "guidances" or recommendations from their regulators that must generally be followed. Some of these laws, regulations, and guidances impose obligations on financial services companies to seek specific protections in their technology contracts. Finally, financial services companies are generally very conservative in their approach to vendor contracting. The foregoing, however, frequently creates significant tension between the need to control risk and the business imperative to "get deals done" and ensure the company's products and services get to market in a timely fashion.

Many technology vendors are unfamiliar with the unique nature and strict regulatory requirements under which financial services companies operate. It frequently comes as a surprise to the vendor that regulators of financial services companies conduct audits of the financial service company's technology contracts to ensure they conform to applicable law and provide the level of protection appropriate for these heavily regulated entities. Because of the foregoing, customers must be ready to explain their unique requirements to their vendors and even provide documentation from their regulators to establish the basis for their concerns.

Three Tools for Better Contracts

We provide three tools for mitigating risk in technology contracts involving financial service companies. First, as in the other chapters in this book, we provide a brief exploration of common key considerations for these types of contracts. Second, at the end of this chapter is a very detailed list of regulatory requirements relevant to technology agreements, including references to applicable regulations, laws, and guidances. The checklist is broken down into general requirements, professional service engagements, software licenses, and offshore engagements.

Finally, in the appendix, we have included a complete copy of the Federal Financial Institutions Examination Council's *Information Technology Examination Handbook* "Outsourcing Technology Services Booklet" (*FFIEC Handbook*). The *FFIEC Handbook* is essentially the guide regulatory auditors use to confirm a financial service company's technology agreements satisfy good contracting practices and mitigate risk to the company and its customers. Everyone involved in negotiating technology agreements for financial services companies should read and understand the handbook. The handbook is also an excellent means of explaining to vendors the regulatory obligations with which the financial services company must comply.

Key Considerations

- First and foremost, negotiators on behalf of the financial services companies must be conversant with their regulatory requirements and be prepared to explain them in plain and simple English to the vendor. The foregoing is one of the most critical points to ensuring negotiations run smoothly and that the goals of the financial services company are achieved. The checklist at the end of this chapter and the *FFIEC Handbook* are two excellent resources for achieving the level of understanding needed to convey the concerns of the financial services company to the vendor.
- Many businesses invest substantial money, in some cases tens of millions of dollars, in developing and maintaining their trademarks and trade names. They are very careful about granting any third party, including their vendors, rights to use those marks and names. Failing to do so may adversely impact their ability to maintain and enforce their rights as trademark owners. In addition to the concerns of any trademark owner, financial services companies are also concerned with use of their names to imply an endorsement of a third party or, worse yet, a recommendation that someone should invest in that third party. Consider a vendor who, without permission, places a statement in its marketing materials that XYZ Broker recommends and endorses its products as the "best in the industry." That statement could be construed as a recommendation by the XYZ Broker to invest in the vendor, resulting in potential regulatory issues for the broker. It is for these reasons that most

financial services companies are extremely strict about granting vendors the right to use their names and marks in customer lists, marketing materials, and case studies. In fact, most include language in their contracts expressly prohibiting the vendor from making any use of their names or marks without the customer's prior written authorization.

- Given the sensitive data involved and the many laws and regulations applicable to highly sensitive financial data, the agreement should include a fully fleshed-out confidentiality clause. That clause should define the types of data at risk, including personally identifiable data, trading information, and potential "insider" information. The contract should make clear the vendor's obligations to hold that data in strict confidence and ensure the information is used solely for performance of the agreement for the financial service company's benefit.
- Most confidentiality clauses include some limitation on their duration (e.g., the obligations of confidentiality will continue for a period of five years after termination). While this may be acceptable in a normal engagement, protection of sensitive financial information, particularly personally identifiable information, should continue in perpetuity. As mentioned in the chapter on nondisclosure agreements, protection of trade secret information should also continue for as long as the information remains protected under applicable law as a trade secret.
- In addition to the various restrictions in the agreement, the vendor should be obligated to comply with the customer's privacy policy with regard to personally identifiable information. The contract should also ensure the vendor is bound by any future updates to the policy. Except in highly unusual situations, the vendor's privacy policy should never take precedence over the policy of a financial services company with regard to its data.
- In the event the vendor breaches security, personally identifiable information is compromised, and a state or federal law would require notice to a consumer, the financial services company should control the content and timing of the notice and be reimbursed by the vendor for the cost of providing the notice. It is also common to require the vendor to reimburse the customer for the cost of providing identity theft insurance to the impacted consumers and the cost of investigating the breach.
- Avoid provisions affording the vendor broad rights to use undefined "aggregated data" (i.e., data that has been cleansed to ensure it cannot be identified to an individual or entity) derived from performance of the agreement. If those rights must be granted, the language should (i) clearly define what aggregation means (in some instances, specific steps to properly cleanse the data of personally identifiable information should be specified, including a clear statement that it is statistically impossible to reidentify any data with an individual); (ii) state that the data is being provided entirely as-is by the financial services company; and (iii) require the vendor to fully indemnify and

hold the financial services company harmless from any and all damages, fines, and costs that may arise out of the vendor's failure to adequately cleanse the data and from *any* use the vendor may make of the data. In some instances, financial services companies may have previously granted exclusive rights to certain forms of aggregated data to others. These obligations should be carefully reviewed before granting rights in aggregated data to others.

- Require the vendor to have conducted background checks on its personnel. In many instances, the customer will also want to conduct its own checks, particularly of on-site workers.
- Depending on the extent and type of data at risk, in addition to the vendor's confidentiality obligations, the contract should specify the vendor's obligations with regard to information security. At its heart this language should require the vendor to comply with, for example, "best industry practices for securing information of the kind provided under the agreement, but in no event less than the level of protection required under all applicable local, state, federal, and international privacy, confidentiality, consumer protection, advertising, electronic mail, data security, data destruction, and other similar laws, rules, and regulations, whether in effect now or in the future." Most financial services companies, however, provide more detailed language. In some instances, exact specifications regarding the level of encryption and firewalls to be employed should be included as part of the contract. At minimum, the contract should address these points:
  - Use of best industry practices, consistent with applicable law (see example language above), to protect the data.
  - A requirement that the vendor promptly report any potential or actual breach of security or confidentiality.
  - A requirement that the vendor furnish log files and other forensic information to assist in the investigation of a breach and to preserve that information.
  - The right of the financial services company to conduct audits of the vendor's security measures and compliance with the agreement.
  - The right for the financial services company to conduct joint security testing, including penetration testing.
  - A requirement that the vendor supply the financial services company with copies of any SSAE 16 or similar audits.
  - Requirements for secure removal/scrubbing of sensitive data from the vendor's systems on termination or expiration of the agreement or from discarded media or in the event the vendor decommissions storage equipment. These requirements are frequently tied to a known data destruction standard (e.g., NIST Special Publication 800-88, Guidelines for Media Sanitization or DoD 5220-22-M Standard).
- Breaches of confidentiality and indemnification obligations should be excluded from all limitations and exclusions of liability.
- In addition to the indemnification rights discussed elsewhere in this book for technology agreements (e.g., an indemnity for intellectual property

infringement), the vendor should also indemnify the financial services company for any damages, fines, sanctions, and liabilities that may arise from the vendor's breach of confidentiality, particularly breaches relating to personally identifiable information.

- As noted above, the financial services company should have the right to audit the vendor with regard to security and contractual compliance. In addition, the financial services company should be able to audit to confirm the accuracy of fees charged. Finally, governmental regulators having authority over the financial services company should have the right to audit and review the vendor.
- In addition to standard termination rights for breach and bankruptcy, the financial services company should have the right to terminate the agreement if one of its regulators identifies the contract and/or the vendor as presenting a material risk that cannot be resolved through negotiations between the parties.
- Some vendor agreements include broad rights of the vendor, and even its agents, to enter and conduct broad audits of the customer. Given the highly regulated nature of a financial services company and the highly sensitive data it controls, these types of rights should generally be rejected. Instead, the customer can offer records for off-site review by the vendor.
- As discussed in the *FFIEC Handbook*, arrangements in which the pricing and purchase of various products are linked to other purchases should be reviewed carefully.
- Use of subcontractors, particularly those who will provide on-site services, have access to sensitive data or are located offshore should be strictly controlled. No subcontractors should be permitted without the customer's express approval. Financial services companies are generally obligated to conduct due diligence of their vendors. Depending on the nature of the subcontractor's involvement, the customer may be under an obligation to also conduct due diligence of the vendor's subcontractors. In some instances (e.g., where the subcontractor will have substantial access to highly sensitive data), it may be appropriate for the financial services company to require the subcontractor to sign a nondisclosure agreement—placing the vendor in direct contractual privity (i.e., they are both parties to the same agreement) with the customer.
  - Beware of vendors who use offshore entities to perform the agreement, particularly if they will be sharing personally identifiable information with those entities. In addition to obvious security and confidentiality concerns, moving data across borders may require consumer consents that may be very difficult to obtain.
- Where the software or service being purchased relates to a regulated activity (e.g., a software platform for processing trades or maintaining account information), the contract should include a warranty that the software or service complies with all applicable laws and that the vendor will furnish, without additional charge, any updates necessary to maintain compliance during the term of the engagement.

Summary

Financial services companies have unique challenges in negotiating technology agreements. In addition to the concerns any customer would have in those types of engagements, financial services companies must also ensure compliance with a wide range of laws, regulations, and guidances from their regulators that directly impact the protections they must require in their technology contracts. To minimize compliance issues, we recommend all financial service companies develop checklists, such as the one at the end of this chapter, to ensure key issues are not overlooked. In addition, negotiators for these entities must be ready to explain and justify these issues to their vendors.

Note: The provisions described in the following checklists typically should be considered and included as described in relevant agreements; however, depending on the facts and circumstances of the services, products, software, service provider, and relationship, discretion and judgment may be required in determining whether to include or modify certain provisions.

The checklist can be used to evaluate and comment on each relevant agreement and retained as a record as to whether and how each issue is addressed.

The first column identifies the regulatory/contractual issue. The second column describes the relevant law, regulation, or guidance as to which the issue relates. The third column is used in evaluating a prospective contract to identify whether a particular requirement is present in the draft agreement. The final column is used to include section references in the proposed agreement that address a particular issue and to record any other issues or comments.

Checklist for Regulatory Considerations in Technology Transactions Involving Financial Services Companies

Provision/Issue	*Regulatory Reference*	*Covered? Yes/No/NA*	*Section Ref./ Comments*
Precontract due diligence. Where relevant and dependent on the level of risk presented, the financial institution should conduct appropriate due diligence of the service provider. Where appropriate, some or all of the following diligence should be conducted: • Experience in implementing and supporting the proposed activity, possibly to include requiring a written proposal • Audited financial statements of the service provider and its significant principals (the analysis should normally be as comprehensive as the financial institution would undertake if extending credit to the party) • Business reputation, complaints, and litigation (by checking references, the Better Business Bureau, state attorneys general offices, state consumers affairs offices, and, when appropriate, audit reports and regulatory reports) • Qualifications, backgrounds, and reputations of company principals, to include criminal background checks, when appropriate • Internal controls environment and audit coverage • Adequacy of management information systems • Business resumption, continuity, recovery, and contingency plans • Technology recovery testing efforts • Cost of development, implementation, and support	Office of Comptroller of Currency ("OCC") 2001-47 Selecting a Service provider and Due Diligence; *FFIEC IT Examination Handbook* (June 2004), Due Diligence, p. 11		

Checklist for Regulatory Considerations in Technology Transactions Involving Financial Services Companies

Provision/Issue	*Regulatory Reference*	*Covered? Yes/No/NA*	*Section Ref./ Comments*
• Reliance on and success in dealing with subcontractors (the financial institution may need to consider whether to conduct similar due-diligence activities for material subcontractors) • Insurance coverage • Business strategies and goals, human resources policies, service philosophies, quality initiatives, and policies for managing costs and improving efficiency • Service provider's culture, values, and business styles should fit the financial institution's environment. • Particular diligence should be exercised with respect to use of offshore service providers. (See the Checklist for Foreign/Offshore Service providers.)			
Reporting. Reports from service providers should be timely, accurate, and comprehensive enough to allow the financial institution to adequately assess performance, service levels, and risks. Discuss frequency and type of reports received (e.g., performance reports, internal controls, control audits, financial statements, security reports, and business resumption testing reports). Consider materiality thresholds and procedures to notify financial institution when service disruptions, security breaches, and other events pose material risk to the financial institution. Consider requiring the service provider to notify in the event of financial difficulty, catastrophic events, material change in strategic goals, and significant staffing changes, all of which could result in a serious impact to service provider service.	OCC 2001-47 Responsibilities for providing and receiving information; *FFIEC Handbook*, Controls, Audit and Reports, p. 14		

Checklist for Regulatory Considerations in Technology Transactions Involving Financial Services Companies

Provision/Issue	*Regulatory Reference*	*Covered? Yes/No/NA*	*Section Ref./ Comments*
Books and records; audit rights. Service provider required to maintain accurate and complete books and records relating to performance; Financial institution must have right to audit service providers (and their subcontractors) as needed to monitor performance under the contract (including OCC). Ensure that periodic independent internal and/or external audits are conducted at intervals and scopes consistent with in-house functions. Include types and frequency of audit reports the financial institution is entitled to receive from the service provider (e.g., financial, internal control, and security reviews). Reserve the right to conduct audits of the function or to engage an independent auditor. Consider whether to accept independent internal audits conducted by the service provider's audit staff or external audits and reviews (e.g., SAS 70 reviews). Audit reports should include a review of service provider's internal control environment relating to the service or product being provided. Reports should also include review of service provider's security program and business continuity program.	OCC 2001-47 The right to audit; *FFIEC Handbook*, Controls, Audit and Reports, p. 14		
OCC supervision. State that performance of services by service provider is subject to OCC examination/oversight and audit by OCC.	OCC 2001-47 OCC Supervision; *FFIEC Handbook*, Controls, p. 14		

Checklist for Regulatory Considerations in Technology Transactions Involving Financial Services Companies

Provision/Issue	*Regulatory Reference*	*Covered? Yes/No/NA*	*Section Ref./ Comments*
Cost/compensation/fees/payment. Fully describe compensation, fees, license fees, and calculations for base services, as well as any charges based upon volume of activity and fees for special requests or services. Indicate which party is responsible for payment of legal, audit, and examination fees associated with the activity. Cost and responsibility for purchasing and maintaining hardware and software. Conditions under which the cost structure can be changed should be addressed in detail, including any limits on any cost increases. Preferable to limit cost escalation and increases to inflation index (e.g., CPI), specified percentage, or service provider's actual increased out-of-pocket costs, as applicable. Applicable taxes should be addressed, including a requirement the service provider assist the financial institution to more accurately determine its tax liability and to minimize such liability to the extent legally permissible. In certain jurisdictions, purely electronic delivery of software and associated documentation may not be subject to sales tax. (See the note below regarding taxes and offshore service providers.)	OCC 2001-47 Cost and compensation; *FFIEC Handbook*, Cost, p. 15, Pricing Methods, p. 17		

Checklist for Regulatory Considerations in Technology Transactions Involving Financial Services Companies

Provision/Issue	*Regulatory Reference*	*Covered? Yes/No/NA*	*Section Ref./ Comments*
Bundling. The vendor may attempt to entice the institution to purchase more than one system, process, or service for a single price—referred to as "bundling." The financial institution should avoid bundled pricing. This practice may result in the financial institution getting a single consolidated bill that may not provide information relating to pricing for each specific system, process, or service. Although the bundled services may appear to be cheaper, the financial institution cannot analyze the costs of the individual services. Bundles may include processes and services that the financial institution does not want or need. It also may not allow the financial institution to discontinue a specific system, process, or service without having to renegotiate the contract for all remaining services.	*FFIEC Handbook*, Bundling, p. 18		
Contract inducements. Financial institutions should not sign servicing contracts that contain provisions or inducements that may adversely affect the financial institution. Such contract provisions may include extended terms (up to ten years), significant increases in costs after the first few years, and/or substantial cancellation penalties. In addition, some service contracts improperly offer inducements that allow the financial institution to retain or increase capital by deferring losses on the disposition of assets or avoiding expense recognition.	*FFIEC Handbook*, Contract Inducement Concerns, p. 19 OCC 2001-47		

Checklist for Regulatory Considerations in Technology Transactions Involving Financial Services Companies

Provision/Issue	*Regulatory Reference*	*Covered? Yes/No/NA*	*Section Ref./ Comments*
Confidentiality and security. Service providers must ensure confidentiality and security of financial institution data and information. Prohibit service provider and agents from using or disclosing financial institution's information, except as necessary to provide the contracted services. If service provider receives nonpublic personal information regarding the financial institution's customers, financial institution should notify service provider to assess the applicability of the privacy regulations, and service provider must implement appropriate security measures designed to meet the objectives of regulatory guidelines with which the financial institution must comply.	Confidentiality and security; *FFIEC Handbook*, Security and Confidentiality, p. 13, Information Security/ Safeguarding, p. 28		
Where appropriate, include specific information security requirements. Financial institutions should require service provider to fully disclose breaches in security resulting in unauthorized intrusions that may materially affect the financial institution or its customers. Service provider should report to the financial institution when material intrusions occur, should estimate the intrusion's effect on the financial institution, and should specify the corrective action taken. Address ability of each party to change security procedures and requirements; changes should remedy any confidentiality/security issues arising out of shared use of facilities owned by the service provider.	OCC 2001-47 Reputation risk; *FIEC Handbook*, Reputation Risk, p. 6		

Checklist for Regulatory Considerations in Technology Transactions Involving Financial Services Companies

Provision/Issue	*Regulatory Reference*	*Covered? Yes/No/NA*	*Section Ref./ Comments*
Publicity. The agreement should prohibit service provider from being able to use the financial institution's name or trademarks in any advertising, promotions, or press releases without the financial institution's prior written consent. **Privacy**. Require service provider to comply with all applicable privacy laws (e.g., G-L-B and FACTA), and financial institution's privacy policies. (See the Checklist for Foreign Offshore Service providers.)	OCC 2001-47 Confidentiality and security; *FFIEC Handbook*, Security and Confidentiality, p. 13; G-L-B; FACTA		
Business resumption, disaster recovery/contingency plans. Provide for continuation of the business function in the event of problems affecting the service provider's operations, including system breakdown and natural (or man-made) disaster. Address service provider's responsibility for backing up and otherwise protecting program and data files, for protecting equipment, and for maintaining disaster recovery and contingency plans. Responsibilities should include testing of the plans and providing results to financial institution. Consider requiring the service provider to provide operating procedures to be carried out in the event business resumption contingency plans are triggered. Include specific timeframes for business resumption and recovery that meet the financial institution's business requirements.	OCC 2001-47 Business resumption and contingency plans; *FFIEC Handbook*, Business Resumption and Contingency Plans, p. 14, Business Continuity Planning, p. 25		

Checklist for Regulatory Considerations in Technology Transactions Involving Financial Services Companies

Provision/Issue	*Regulatory Reference*	*Covered? Yes/No/NA*	*Section Ref./ Comments*
Indemnification. Service provider should be required to defend, indemnify, and hold the financial institution harmless from third-party claims for intellectual property infringement, and claims arising out service provider's alleged negligence, breach of confidentiality (particularly with regard to customer information), and violation of law. In the event financial institution indemnifies service provider, financial institution should not be liable for claims arising out of any acts, omissions, or failure of the service provider.	OCC 2001-47 Indemnification; *FFIEC Handbook*, Indemnification, p. 16		
Insurance. The service provider should maintain adequate insurance and should notify the financial institution of material changes to coverage. Insurance requirements should reflect the level of risk presented by the proposed engagement.	OCC 2001-47 Insurance; *FFIEC Handbook*, Controls, p. 14		
Dispute resolution. Consider whether to provide for a dispute resolution process (arbitration, mediation, or other means) for the purpose of resolving problems between the financial institution and service provider in an expeditious manner, and whether it should provide that the service provider continue to perform during the dispute resolution period.	OCC 2001-47 Dispute resolution; *FFIEC Handbook*, Dispute Resolution, p. 16		

Checklist for Regulatory Considerations in Technology Transactions Involving Financial Services Companies

Provision/Issue	*Regulatory Reference*	*Covered? Yes/No/NA*	*Section Ref./ Comments*
Limitations/disclaimer of liability/damages. Most standard service provider contracts limit the service provider's liability. The financial institution should determine whether the proposed limit is in proper proportion to the amount of loss the financial institution might experience as a result of the service provider's failure to perform. A waiver of consequential damages is acceptable provided there is a carve-out for indemnification claims and breach of confidentiality/ security. Negotiations for a cap on damages should start at no less than two times all fees paid and financial institution should not agree to less than twelve months of fees paid during the prior year without senior management approval; and cap on damages should include the same carve-out for indemnification claims and breach of confidentiality/ security. Limitations of liability should be drafted to provide mutual protection for both parties. Avoid limitations of liability that protect only the service provider.	OCC 2001-47 Limits on liability; *FFIEC Handbook*, Limitation of Liability, p. 16		

Checklist for Regulatory Considerations in Technology Transactions Involving Financial Services Companies

Provision/Issue	*Regulatory Reference*	*Covered? Yes/No/NA*	*Section Ref./ Comments*
Term/termination/default. Significant risks are associated with contract default and/or termination. Stipulate clearly what constitutes default; identify remedies; allow for opportunities to cure defaults (typically thirty days). Extent and flexibility of termination rights sought may vary with the type of service. Termination rights for change in control, merger or acquisition, convenience, substantial increase in cost, repeated failure to meet service standards, failure to provide critical services and required notices, failure to prevent violations of law or unfair and deceptive practices, bankruptcy, company closure, and insolvency. Consider renewal terms and appropriate length of time required to avoid automatic renewal. Provide for financial institution's right to terminate upon reasonable notice and without penalty in the event that the OCC formally objects to the particular service provider relationship. Allow for financial institution to terminate the relationship in a timely manner without prohibitive expense. Provide timeframes to allow for transition assistance and the orderly conversion to another service provider; and timely return of the financial institution's data and other financial institution resources. Any costs and service provider's obligations associated with transition assistance should be clearly stated (e.g., service provider expense if termination due to service provider default).	OCC 2001-47 Default and termination; *FFIEC Handbook*, Duration, p. 15, Termination, p. 16		

Checklist for Regulatory Considerations in Technology Transactions Involving Financial Services Companies

Provision/Issue	*Regulatory Reference*	*Covered? Yes/No/NA*	*Section Ref./ Comments*
Customer complaints. Where relevant, service provider required to forward any complaints it receives from the financial institution's customers. Specify whether the financial institution or service provider is responsible for responding to the complaints. If the service provider responds, a copy of the response should be forwarded to the financial institution.	OCC 2001-47 Customer complaints		
Background checks. Provide for financial institution's right, at its discretion, to require full criminal, employment, and/or drug background checks on all service provider personnel performing services for the financial institution. (Consider whether the checks should be conducted prior to engaging service provider.) Background checks are particularly relevant in offshore engagements. **Assignment**. Service provider should not be permitted to assign the agreement without financial institution's consent. Financial institution should be able to freely assign to an affiliate or in the event of a sale, merger, acquisition, change of control.	OCC 2001-47 Selecting a Service provider and Due Diligence; *FFIEC Handbook*, Due Diligence, p. 11		
Miscellaneous. Consider appropriate miscellaneous provisions to include, such as governing law and venue in an acceptable jurisdiction, survival, integration/entire agreement, modifications only in writing, and inapplicability of click-wrap/web-based terms, conditions, licenses, and disclaimers.	OCC 2001-47 Scope of arrangement; *FFIEC Handbook*, Assignment, p. 16		

Checklist for Professional Services Agreements

Provision/Issue	*Regulatory Reference*	*Covered? Yes/No/NA*	*Section Ref./ Comments*
Description of services and/or products. Clearly identify the scope of services and/or products to be provided, including the frequency, content, and format of the services or products to be provided. Where applicable, reference statement of work, schedule, milestones, and deliverables.	OCC 2001-47 Scope of arrangement; *FFIEC Handbook*, Scope of Services, p. 13		
Support services. Identify software support and maintenance, training of employees, and customer service.	OCC 2001-47 Scope of arrangement; *FFIEC Handbook*, Scope of Service, p. 13		
Third-party software. Identify third-party software required to use the services and responsibility for acquiring and fees. Identify or prohibit, as applicable, open-source software supplied by service provider. Require identification of all applicable third-party terms and conditions.			
Permitted activities; financial institution premises. Describe which activities service provider is permitted to conduct, whether on or off the financial institution's premises, and describe the terms governing the use of the financial institution's space, personnel, and equipment.	OCC 2001-47 Scope of arrangement		
Joint responsibilities. When financial institution and service provider employees are used jointly, their duties and responsibilities should be clearly articulated.	OCC 2001-47 Scope of arrangement; *FFIEC Handbook*, Scope of Services, p. 13		

Checklist for Professional Services Agreements

Provision/Issue	*Regulatory Reference*	*Covered? Yes/No/NA*	*Section Ref./ Comments*
Subcontracting. Indicate whether the service provider is prohibited from assigning any portions of the contract to subcontractors or other entities. Preferable to prohibit subcontracting core or critical services, and maintaining control and approval rights over permitted subcontracting. Subcontracting services that involve access to customer information should be strictly controlled. See the Offshore Checklist for subcontracting to offshore service providers. Prohibit subcontracting to offshore service providers where appropriate.	OCC 2001-47 Scope of arrangement; *FFIEC Handbook*, Sub-contracting and Multiple Service Provider Relationships, p. 15		
Testing and acceptance. Include provision for testing and acceptance by financial institution, including procedure for determining acceptance criteria, and service provider obligation to remedy failures, refund for failed implementation.	OCC 2001-47 Performance measures or benchmarks; *FFIEC Handbook*, Performance Standards, p. 13, Controls, p. 14, Regulatory Compliance, p. 16		
Warranties. Include appropriate service provider warranties such as services warranty (professional, competent, trained employees); compliance with all applicable laws and regulatory requirements; authority to enter into agreement; conformance to specifications; and noninfringement.			

Checklist for Professional Services Agreements

Provision/Issue	*Regulatory Reference*	*Covered? Yes/No/NA*	*Section Ref./ Comments*
Service levels/performance measures. Performance measures should define the expectations and responsibilities for both parties. Requirements should be sufficient to enable effective monitoring of ongoing service provider performance and success of the arrangement; used to motivate third-party performance, especially if poor performance is penalized or outstanding performance rewarded. Industry standards should provide a reference point for commodity-like services, such as payroll processing. Use agreed-upon range of measures for customized services. Ensure service provider has an obligation to report service level compliance. Set and monitor parameters for financial functions, including payments processing or extensions of credit on behalf of the financial institution.	OCC 2001-47 Performance measures or benchmarks; *FFIEC Handbook*, Performance Standards, p. 13, Controls, p. 14, Service Levels Agreements, p. 17		
Change control. Include provisions governing provision of services, scope of work, and compensation for services outside the original scope of the agreement, including changes to systems, controls, key project personnel, and service locations.	OCC 2001-47 Scope of arrangement; *FFIEC Handbook*, Controls, p. 14		
Staffing; employment issues. Allow for reasonable financial institution control over approval and replacement of service provider staffing, particular for personnel on site; consider limits on reassignment of key personnel; require service provider to promptly notify financial institution and replace personnel where necessary and to accomplish continuity of services.			

Checklist for Professional Services Agreements

Provision/Issue	*Regulatory Reference*	*Covered? Yes/No/NA*	*Section Ref./ Comments*
Include language making clear the financial institution will not be deemed the employer of any service provider personnel and that the service provider will indemnify and hold the financial institution harmless from any employment-related claims. The service provider is solely responsible for payment of all employment-related taxes.	OCC 2001-47 Scope of arrangement; *FFIEC Handbook*, Controls, p. 14		
Intellectual property ownership/license. State whether and how service provider has the right to use the financial institution's data, hardware and software, system documentation, and intellectual property, such as financial institution's name, logo, trademark, and copyrighted materials. Any use of financial institution property should be subject to a clearly worded license. Indicate whether any records generated by service provider are the property of the financial institution. Consider whether developed software and other intellectual property should be owned by the financial institution, or if not, financial institution granted exclusivity for a period of time. If not owned by financial institution, ensure the financial institution has an appropriate license to use the developments. If licensing software, consider establishing escrow agreements to provide for financial institution's access to source code and programs under certain conditions (e.g., insolvency of the service provider), documentation of programming and systems, and verification of updated source code.	OCC 2001-47 Ownership and license; *FFIEC Handbook*, Ownership and License, p. 15		

Checklist for Software License Agreements

Provision/Issue	*Regulatory Reference*	*Covered? Yes/No/NA*	*Section Ref./ Comments*
Scope of license. Clearly identify the software and scope of license rights being granted; ability of financial institution to use contractors and outsourcers; and ability of affiliates and divested entities to use software.	OCC 2001-47 Scope of arrangement; *FFIEC Handbook*, Scope of Services, p. 13		
Support services. Identify software support and maintenance, error corrections, telephone support/help desk, updates, modifications, training of employees, customer service, and backups. Consider minimum period of time (e.g., five years) when service provider will be required to support if support is renewed by financial institution.	OCC 2001-47 Scope of arrangement; *FFIEC Handbook*, Scope of Services, p. 13, Duration, p. 15		
Third-party software. Identify third-party software required to use the service provider software, responsibility for acquiring and fees. Identify or prohibit, as applicable, open-source software. Require identification of all applicable third-party terms and conditions.	OCC 2001-47 Scope of arrangement		
Permitted activities; financial institution premises. Describe which activities service provider is permitted to conduct, whether on or off the financial institution's premises, and describe the terms governing the use of the financial institution's space, personnel, and equipment.			

Checklist for Software License Agreements

Provision/Issue	*Regulatory Reference*	*Covered? Yes/No/NA*	*Section Ref./ Comments*
Joint responsibilities. When financial institution and service provider employees are used jointly, their duties and responsibilities should be clearly articulated.	OCC 2001-47 Scope of arrangement		
Subcontracting. Indicate whether the service provider is prohibited from assigning any portions of the contract to subcontractors or other entities. Preferable to prohibit subcontracting core or critical services, and maintaining control and approval rights over permitted subcontracting. Subcontracting services that involve access to customer information should be strictly controlled. (See the Offshore Checklist for subcontracting to offshore service providers. Prohibit subcontracting to offshore service providers where appropriate.)	OCC 2001-47 Scope of arrangement; *FFIEC Handbook*, Sub-contracting and Multiple Service Provider Relationships, p. 15		
Testing and acceptance. Include provision for testing and acceptance by financial institution, including procedure for determining acceptance criteria, service provider obligation to remedy failures, and refund for failed implementation.			
Warranties. Include appropriate service provider warranties such as software will conform to specifications, services warranty (professional, competent, trained employees); compliance with laws; authority to enter into agreement; noninfringement; no viruses/destructive code; DST compliance; and no known performance issues.			

Checklist for Software License Agreements

Provision/Issue	*Regulatory Reference*	*Covered? Yes/No/NA*	*Section Ref./ Comments*
Service levels/performance measures. Performance measures should define the expectations and responsibilities for both parties. Requirements should be sufficient to enable effective monitoring of ongoing service provider performance and success of the arrangement, and should be used to motivate third-party performance, especially if poor performance is penalized or outstanding performance rewarded. Consider service levels for availability of software (particular if service provider hosted), problem call response times, and time to repair. Ensure service provider has an obligation to report service level compliance.	OCC 2001-47 Performance measures or benchmarks; *FFIEC Handbook*, Performance Standards, p. 13, Service Level Agreements, p. 17		
Intellectual property ownership/license/modifications/escrow. State whether and how service provider has the right to use the financial institution's data, hardware and software, system documentation, and intellectual property, such as financial institution's name, logo, trademark, and copyrighted material. Any use of financial institution property should be subject to a clearly worded license. Indicate whether any records generated by service provider are the property of the financial institution. Consider whether financial institution will obtain any source code and/or be permitted to make modifications and ownership of such modifications. Consider establishing escrow agreements to provide for financial institution's access to source code and programs under certain conditions (e.g., insolvency of the service provider), documentation of programming and systems, and verification of updated source code.	OCC 2001-47 Ownership and license; *FFIEC Handbook*, Ownership and license, p. 15		

Checklist for Foreign/Offshore Service Providers

Provision/Issue	*Regulatory Reference*	*Covered? Yes/No/NA*	*Section Ref./ Comments*
Precontract due diligence. In addition to due diligence appropriate for a domestic service providers (see Checklist for General Provisions/All Service providers), the financial institution should engage in even higher due diligence prior to engaging an offshore service provider. The due diligence process should include an evaluation of the foreign-based service provider's ability—operationally, financially, and legally—to meet the financial institution's servicing needs given the foreign jurisdiction's laws, regulatory requirements, local business practices, accounting standards, and legal environment. The due diligence also should consider the parties' respective responsibilities in the event of any regulatory changes in the United States or the foreign country that could impede the ability of the financial institution or service provider to fulfill the contract.	OCC 2002-16 Due Diligence; *FFIEC Handbook*, Appendix C, Due Diligence, p. C-3		
Privacy. Pay special attention to protecting privacy of customers and the confidentiality of financial institution records given US law and the foreign jurisdiction's legal environment and regulatory requirements. Appropriate warranties, confidentiality provisions, and information security requirements should be included in the agreement.	OCC 2002-16; *FFIEC Handbook*, App. C, Security, Confidentiality and Ownership of Data, p. C-3		

Checklist for Foreign/Offshore Service Providers

Provision/Issue	*Regulatory Reference*	*Covered? Yes/No/NA*	*Section Ref./ Comments*
Choice of law. Carefully consider which country's law should control the relationship and insert appropriate choice of law and jurisdictional language that provides for resolution of all disputes between the parties under the laws of the acceptable jurisdiction. (*Note: Choice of law and jurisdictional provisions help to ensure continuity of service, to maintain access to data, and to protect nonpublic customer information. The provisions, however, can be subject to interpretation of foreign courts relying on local laws, which may substantially differ from US laws in how they apply and enforce choice of law covenants, what they require of financial institutions, and how they protect financial institution customers. As part of due diligence process, financial institution should obtain legal review from counsel experienced in that country's laws regarding the enforceability of all aspects of the subject contract and any other legal ramifications.*) **Taxes**. Identify any applicable local taxes. Address responsibility for changes in local taxes (e.g., service taxes in outsourcing engagements) that may occur during the term of the agreement.	OCC 2002-16 Choice of law; *FFIEC Handbook*, App. C, Choice of Law, p. C-4		
Confidentiality. Ensure service provider is prohibited from disclosing or using financial institution data or information for any purpose other than to carry out the contracted services. All information shared by the financial institution with service provider, regardless of how the service provider processes, stores, copies, or otherwise reproduces it, remains			

Checklist for Foreign/Offshore Service Providers

Provision/Issue	*Regulatory Reference*	*Covered? Yes/No/NA*	*Section Ref./ Comments*
solely the property of the financial institution. Require service provider to implement security measures that are designed to safeguard customer information. (*Note: Sharing of nonpublic customer-related information from US offices with a foreign-based service provider must comply with the OCC's privacy regulation, including requisite disclosures to and agreements with customers who would be affected by the financial institution's relationship with the service provider. The financial institution should not share any nonpublic OCC information, such as an examination report, with a foreign-based service provider except with express OCC approval. Such nonpublic OCC information remains the OCC's property, and the financial institution should take all required measures to protect the information's confidentiality.*)	OCC 2002-16 Confidentiality of Information; *FFIEC Handbook*, Security and Confidentiality, p. 13, *FFIEC Handbook*, App. C, Security, Confidentiality and Ownership of Data, p. C-3		
Financial institution access to information. Critical data or other information related to services provided by a foreign-based service provider must be readily available at the financial institution's US office(s). Such information should include copies of contracts, due diligence, and oversight and audit reports. In addition, the financial institution should have an appropriate contingency plan to ensure continued access to critical information and service continuity and resumption in the event of unexpected disruptions or restrictions in service resulting from transaction, financial, or country risk developments.	OCC 2002-16 Access to Information; *FFIEC Handbook*, App. C, Regulatory Authority, p. C-4, Regulatory Agency Access to Information, p. C-5		

Checklist for Foreign/Offshore Service Providers

Provision/Issue	*Regulatory Reference*	*Covered? Yes/No/NA*	*Section Ref./ Comments*
OCC access to information. Use of a foreign-based service provider and the location of critical data and processes outside US territory must not compromise the OCC's ability to examine the financial institution's operations. Agreement should establish the relationship in a way that permits and does not diminish the OCC's access to data or information needed to supervise the financial institution. The financial institution should not outsource any of its information or transaction processing to third-party service providers that are located in jurisdictions where the OCC's full and complete access to data or other information may be impeded by legal, regulatory, or administrative restrictions unless copies of all critical records also are maintained at the financial institution's US offices. Copies of the results of the financial institution's due diligence efforts and regular risk management oversight, performance and audit reports on the foreign-based third-party service provider, as well as all policies, procedures, and other important documentation relating to the financial institution's relationship with the service provider, should be maintained in English for review by examiners at the financial institution's office(s).	OCC 2002-16 Access to Information; *FFIEC Handbook*, App. C, Regulatory Authority, p. C-4, Regulatory Agency Access to Information, p. C-5		

Chapter 21

Source Code Escrow Agreements

CHECKLIST

What Type of Escrow?

- ☐ Two-party
- ☐ Three-party
- ☐ Self-escrow

Release Conditions

- ☐ Insolvency of vendor
- ☐ Filing of voluntary or involuntary bankruptcy proceedings that remain undismissed
- ☐ General assignment for the benefit of creditors
- ☐ Ceasing to provide support and maintenance services
- ☐ Breach of the license agreement

Key Issues to Be Addressed in Software License

- ☐ Requirement that source code be escrowed
- ☐ Identify approved escrow agent
- ☐ Definition of "source code"

Identify release conditions

- ☐ Require vendor to update the source code to reflect current version of the software
- ☐ Identify all relevant fees and the responsible party
- ☐ Include right to request verification services from the escrow
  - – Include cost shifting to vendor if verification fails
- ☐ On the occurrence of a release condition, include right for customer to use and modify the software to provide its own support

Techniques

- ☐ In critical applications, include right for the customer to use a consultant to verify the source code is well written and documented.
  - – Nondisclosure agreement with consultant
  - – Termination right for customer in the event of adverse findings
- ☐ Consider requiring the vendor to deposit names and contact information of key developers in the escrow
- ☐ Ensure the escrow company is well established and financial viable

Overview

Whenever a customer enters into a license agreement for software, it takes the risk that the vendor may go out of business, file bankruptcy, or, simply, cease providing support. Depending on how critical the software is to the customer, these events could be catastrophic, stranding the customer with a piece of software for which it has no means of support. If the investment in licensing the software is substantial (e.g., hundreds of thousands or, even, millions of dollars in fees, protracted implementation), the customer must have some means of protecting itself. While conducting vendor due diligence—particularly with regard to financial wherewithal—is clearly one of the most important protections, careful customers also generally require the vendor to "escrow" the source code for its software. In this chapter, we discuss what "escrowing" means, the types of escrow arrangements, the real-world benefits of escrowing, and common issues.

While this chapter focuses on the use of source code escrows, we should point out that another option may be available in a limited number of instances. That option is a service some escrow companies are now offering that has nothing to do with traditional source code escrow. Rather, the escrow company enters into an agreement with the vendor to fully implement a copy of its software on servers hosted by the escrow company. In the event the vendor ceases to do business, files bankruptcy, etc., the customer can obtain instant access to the version already being hosted by the escrow company. This is not a perfect solution. It is only available in limited instances, and the customer must accept very limited contractual protections from the escrow company. It is, at best, a temporary solution to be used while transitioning to a replacement vendor.

What Does It Mean to Escrow Source Code?

Almost all commercial software is licensed so that the customer only receives the object code version of the application. The object code is the version of the software that can be read and understood by a computer, but not by humans. To modify the software, fix bugs, and otherwise maintain and support the application, one must have the source code (i.e., the version of the programming that can be understood and edited by humans). The source code is generally considered a very valuable trade secret of the vendor and carefully protected. In most instances, no customer is ever granted access to the source code.

As mentioned in the introduction, there may be circumstances in which the customer must have access to the source code or face significant harm (e.g., if the vendor declares bankruptcy or ceases to support the software). To protect against those circumstances, the concept of a "source code escrow" was created. A source code escrow is, basically, an arrangement where the vendor entrusts its source code to a third-party escrow agent who holds the source code in trust until a "release condition" occurs (e.g., the vendor files for bankruptcy protection, or the vendor ceases to do business). In the event of a release condition, the escrow agent is contractually bound to furnish a copy of the source code to the customer. If the vendor disputes the existence of a release condition, the escrow agent will hold the source code until a dispute resolution process is completed. That process can take many weeks or, even, months.

Professional escrow companies have their own form agreements for their engagements. They require their customers to use those agreements and, except for release conditions, those contracts are generally nonnegotiable. It is very rare for an escrow company to agree to use contracts drafted by third parties.

Types of Escrow Agreements

There are basically two types of escrow agreements: two-party and three-party. In a two-party engagement, the vendor and escrow agent sign the escrow agreement. The customer is not a direct party to the contract, but signs a separate "beneficiary form," which gives the customer the right to receive the source code on the occurrence of a release condition (as discussed below). Two-party agreements have the following characteristics:

- They are generally used when a vendor signs a single escrow agreement for all of its customers.
- Each customer signs a separate beneficiary form.
- Customers may be required to pay a relatively small fee to become a beneficiary. There may also be certain copying and other fees if a release condition occurs.
- The escrow agreement is presented to customers as essentially nonnegotiable, forcing customers to accept the all-important release conditions previously negotiated by the vendor and escrow agent.
- Two-party agreements are more frequently found in smaller transactions and those involving software that does not require extensive customer customizations.

In three-party escrow arrangements, the customer, vendor, and escrow agent are all direct parties to the contract. Three-party agreements have the following characteristics:

- Frequently found in larger transactions, particularly those involving customized software.
- Fees to be paid by the customer are generally higher because the vendor is required to enter into an entirely new agreement for each customer who desires an escrow arrangement.
- The agreement, particularly the release conditions, is capable of negotiation.

When talking about whether and to what extent an escrow agreement may be negotiated, it must be understood that, absent very unusual circumstances (e.g., a very large software vendor who does significant business with the escrow company and, therefore, has greater leverage to modify contract terms), the terms of the escrow agreement that relate to liability and other obligations of the escrow agent are generally very difficult, if not impossible, to negotiated. This is particularly true of large, well-established escrow companies. These largely nonnegotiable terms relating to the escrow agent's liability and other obligations, however, should be contrasted with provisions relating to the release conditions for the source code. Those conditions can be freely negotiated. Whatever the vendor and customer agree upon, the escrow agent is generally willing to include in the escrow agreement. For ease of revision and to differentiate those terms from the rest of the agreement, release conditions are frequently set forth in an exhibit to the contract.

While the two- and three-party escrow arrangements described above are far and away the most common, one other arrangement is also possible: the self-escrow. As the name suggests, in a self-escrow arrangement, the vendor delivers a copy of the source code to the customer when the software license agreement is signed. The agreement specifies that the customer may not use the source code unless a release condition occurs. In the meantime, the customer must keep the source locked in a secure container, usually a safe. The advantage of a self-escrow is that the customer has immediate access to the source code if a release condition occurs. In normal escrow arrangements using an escrow agent, there can be many weeks of delay in obtaining the source code. That timing may be essential in some business critical applications.

Self-escrow arrangements are rare, but they should be considered for highly critical applications or in situations involving smaller vendors who do not have an established relationship with a source code escrow agent.

Release Conditions

Every source code escrow agreement includes a list of release conditions. On the occurrence of any of those conditions, unless the vendor is disputing the occurrence, the escrow agent is contractually bound to furnish a copy of the source code to the customer. Since the release conditions will result in the customer obtaining a copy of the vendor's most valuable asset (the source code), vendors are very reticent to include any but the most restrictive release conditions. Common conditions include:

- Insolvency of the vendor.
- The vendor is subject to voluntary or involuntary bankruptcy and fails to have the matter dismissed within thirty (30) days.

These conditions are frequently supplemented to add more robust protection for the customer:

- The making of a general assignment by the vendor for the benefit of its creditors
- In the event the vendor ceases to maintain and support the licensed software for reasons other than the customer's failure to pay for, or election not to receive, maintenance and support services
- Termination of the license agreement for breach by the vendor

Key Issues for Escrow Agreements

This section presents key considerations and risks presented by escrow agreements.

The software license agreement with the vendor should clearly identify the following obligations and rights for the escrow:

- The vendor has an obligation to establish and maintain the escrow during the term of the agreement. For perpetual licenses, it is not uncommon to limit the obligation to the period in which support is purchased and, perhaps, a year or two thereafter.

- The software and all updates and bug fixes must be promptly deposited on their general availability. At minimum, the deposit should be updated on a quarterly basis.
- The vendor contract should specify the party responsible for the fees for establishing and maintaining the escrow.
- The vendor contract should expressly permit the customer to request verification services from the escrow agent. For a fee, the escrow agent will compile the deposited code and compare it with the version of the object code currently being licensed by the vendor to ensure they match (i.e., that the deposit isn't out of date or, worse yet, a blank CD with nothing on it). If the verification is correct, the cost of the escrow agent's services is the responsibility of the customer. However, if the verification shows the vendor has not fulfilled its obligation to ensure the deposit is up to date, the cost of the verification services should shift to the vendor. This is a key protection to ensure the vendor has the incentive to keep the deposit current.
- Reject arguments by the vendor that verification services are unnecessary in that if they fail to keep the deposit current, they will be in breach of contract. The problem with that argument is that when the customer needs the source code the most (e.g., when the vendor has filed for bankruptcy protection), it will be too late to seek damages.
  - In the event a release condition occurs, the vendor contract should make clear that the customer will have the right to modify the source code and otherwise use it to provide its own maintenance and support.
  - Vendors sometimes include language that if they "cure" whatever conditions resulted in the release of source code, the customer must return the source code. While this seems fair, it actually presents a significant problem. If a release occurs, the customer will likely spend thousands, if not tens of thousands, of dollars to review and understand the source code and prepare to provide its own support. If the customer knows it may have to return the source code at any time if the release condition is cured, it will be very reticent to make that investment greatly diminishing the value of the escrow.
  - In certain highly critical applications, it may be appropriate to include in the software license a right for the customer to engage a third-party consultant to assess and evaluate the source code to determine if it is well written and documented (i.e., that it is not "spaghetti code"—code that was not developed using accepted programming practices and that is poorly written and documented). In these situations, the customer would have no access to the source code. Only the third-party consultant would have access. The consultant would typically sign a nondisclosure agreement (NDA) directly with the vendor to ensure the source code remains confidential.

- Frequently, this right is exercised at the outset of the agreement. If the consultant issues an adverse report, the customer should have a termination right.
- All release conditions should be identified:
    - Bankruptcy or insolvency of the vendor
    - Vendor ceasing to do business
    - Vendor abandoning its support and maintenance obligations
    - Breach of the agreement by vendor

    This last release condition is generally the most difficult to achieve. It is, however, an excellent incentive for the vendor to avoid breaching the contract and, therefore, a strong protection for the customer.
- The vendor contract should include a statement that it is amending the underlying escrow agreement with regard to the release conditions.
- Include a detailed definition of what constitutes "source code." That is, source code is not just the software, itself, but related files necessary for the code to work properly and all related programming documentation. A typical definition may include "the source code of the licensed software and all related compiler command files, build scripts, scripts relating to the operation and maintenance of such application, application programming interface (API), graphical user interface (GUI), object libraries, all relevant instructions on building the object code of such application, and all documentation relating to the foregoing, such that collectively the foregoing will be sufficient to enable a person possessing reasonable skill and expertise in computer software and information technology to build, load- and operate the machine-executable object code of such application, to maintain and support such application and to effectively use all functions and features of the licensed software."
- In some transactions, particularly those involving small vendors, it may be appropriate to require the vendor to include in its escrow deposit the names and contact information for all key development personnel. This will allow the customer to potentially locate and hire (either as an employee or as a contractor) one or more of the developers to assist the customer in understanding and using the source code.

■ Always make sure to use a recognized escrow company that specializes in source code escrow engagements.
- Avoid small companies who lack appropriate financial wherewithal and may, themselves, go out of business.
- Avoid banks, lawyers, and others who offer source code escrow services as a side business.
- Obtain and review the escrow agent's form agreements. They should be readily available on their website or on request.

- Identify whether a two- or three-party agreement will be used. In most cases, unless the customer is willing to pay the additional fees to establish a new escrow arrangement, the customer must use the vendor's existing escrow agreement.
    - Three-party agreements are generally preferred because they afford the customer potentially greater ability to negotiate release conditions. However, because they require a new agreement for each customer, the customer may be required to pay higher fees to establish the escrow.
    - In two-party arrangements, the vendor generally portrays them as nonnegotiable, offering the same uniform release conditions and other terms to all of its customers.

- Identify any fees to be paid by the customer/beneficiary.
  - In general, in two-party agreements, the customer should decline to pay any fees associated with establishing and maintaining the escrow. If pressed, the customer should only commit to paying the fees associated with becoming a beneficiary and any copying costs associated with a release of the source code.
- In three-party agreements, depending on the size of the engagement, many vendors will bear the escrow's costs. However, it is not uncommon for the parties to equally share the costs.

Summary

An escrow agreement should not be looked on as a panacea. The code could be so poorly written and documented that it could take months to understand its operation. The cost to the customer of understanding the source code and providing its own support may well exceed the cost of simply licensing and implementing a replacement product.

Escrows can provide important additional protection to customers, particularly when the licensor is small, financially unstable, or when a critical application is involved. Customers, however, should understand the realities of actually trying to provide their own support using what could be hundreds of thousands of lines of code written by a third party and for which there may be little or no documentation. In addition, in order to make the code work, the customer may have to purchase numerous third-party license agreements for embedded software within the licensor's original code. Finally, for cloud engagements, the code is set up for a multi-tenant hosted environment. If a customer gains access to the source code for a cloud solution, implementing it on its own systems for a single user may prove very difficult.

Chapter 22

Integrating Information Security into the Contracting Life Cycle

CHECKLIST

Use the Three Tools for Better Integrating Information Security into the Contract Life Cycle

- ☐ Precontract due diligence
- ☐ Key contractual protections
- ☐ Information security requirements exhibit

Precontract Due Diligence

- ☐ Develop a form due diligence questionnaire
- ☐ Ensure the questionnaire covers all key areas
- ☐ Use the questionnaire as an early means of identifying security issues
- ☐ Use the questionnaire to conduct an "apples-to-apples" comparison of prospective vendors

Key Contractual Protections

- ☐ Fully fleshed-out confidentiality clause
- ☐ Warranties
  - Compliance with best industry practices; specify the relevant industry
  - Compliance with applicable laws and regulations (e.g., HIPAA, GLB)
  - Compliance with third-party standards (e.g., payment card industry, data security standard, payment application data security standard)
  - Compliance with customer's privacy policy
  - Prohibition against making data available offshore
  - Responses to due diligence questionnaire are true and correct
- ☐ General security obligations
  - All reasonable measures to secure and defend systems
  - Use of industry standard antivirus software
  - Vulnerability testing
  - Immediate reporting of actual or suspected breaches
  - Participation in joint audits
  - Participation in regulatory reviews
- ☐ Indemnity against claims, damages, costs arising from a breach of security
- ☐ Responsibility for costs associated with providing breach notifications to consumers; control of timing and content of notice
- ☐ Forensic assistance
  - Duty to preserve evidence
  - Duty to cooperate in investigations
  - Duty to share information
- ☐ Audit rights
  - Periodic audits to confirm compliance with the agreement and applicable law
  - Provision of any appropriate SSAE 16 (now SSAE 18, often referred to as a SOC 1 audit), SOC 2 audit, ISO/IEC 27001, or similar audits
- ☐ Limitation of liability should exclude breaches of confidentiality from all limitations and exclusions of liability
- ☐ Post-contract policing

Information Security Requirements Exhibit

- ☐ Where appropriate, develop an exhibit, statement of work, or other contract attachment describing specific required information security measures
- ☐ Use of wireless networks
- ☐ Removable media
- ☐ Encryption
- ☐ Firewalls
- ☐ Physical security

Overview

Newspapers and trade journals feature a growing number of stories detailing instances in which organizations have entrusted their most sensitive information and data to a vendor only to see that information compromised because the vendor failed to implement appropriate information security safeguards. Worse yet, those same organizations are frequently found to have performed little or no due diligence regarding their vendors and have failed to adequately address information security in their vendor contracts, in many instances leaving the organizations without a meaningful remedy for the substantial harm they have suffered as a result of a compromise. That harm may take a variety of forms: damage to business reputation, loss of business, potential liability to the data subjects, and regulatory and compliance issues.

Whether the information at issue is highly regulated data identifiable to individuals (e.g., nonpublic financial information, protected health information, or the myriad of other information now subject to state, federal, and international protection relating to individuals) or sensitive business information, including trade secrets and other proprietary information, companies must ensure that information is adequately protected by their vendors. This chapter discusses three tools companies may use to reduce information security threats posed by their vendor relationships, to ensure proper due diligence is conducted and documented, and to provide remedies in the event of a compromise. Those tools are: (i) the due diligence questionnaire; (ii) key contractual protections; and (iii) the use in appropriate circumstances of an information security requirements exhibit. Whenever a vendor will have access to an organization's network, facilities, personal data, or other sensitive or valuable data, one or more of these tools should be used.

By implementing these measures, the company can better integrate information security into the entire contracting process—as opposed to simply having it be a "bolt-on" at the time of contract negotiations.

Due Diligence: The First Tool

Companies should conduct some form of due diligence before entrusting vendors with sensitive information or with access to their systems. Unfortunately, most companies conduct this review on an ad hoc basis, informally, without clear documentation. In very few instances is the outcome of that due diligence actually incorporated into the parties' contract. This approach to due diligence may no longer be appropriate or reasonable in the context of today's business and regulatory environment. To help ensure proper documentation and uniformity of the due diligence process, especially for high-risk arrangements, companies should consider developing a standard due diligence questionnaire or adopting an industry standard one for the company's industry for prospective vendors to complete.

Key Contractual Protections: The Second Tool

In the overwhelming majority of engagements, the underlying contract entered into between a company and its vendor will have little or no specific language relating to information security. At most, there is a passing reference to undefined security requirements set forth in the vendor's "then-current security policy" and the inclusion of a basic confidentiality clause. Today's best practices in vendor contracting suggest far more specific language is required, particularly when regulated personally identifiable information is at risk. The following protections should be considered for inclusion in relevant vendor contracts:

- **Confidentiality.** A fully fleshed-out confidentiality clause should be the cornerstone for information security protections in every agreement. The confidentiality clause should be broadly drafted to include all information the company desires to be held in confidence. Specific examples of protected information should be included (e.g., source code, marketing plans, new product information, trade secrets, financial information, and personally identifiable information). While the term of confidentiality protection may be fixed (for, say, five years), ongoing, perpetual protection should be expressly provided for consumer information and trade secrets of the business. Requirements that the company mark relevant information as "confidential" or "proprietary" should be strictly avoided. These types of requirements are unrealistic in the context of most vendor relationships. The parties frequently neglect to comply with these requirements, resulting in proprietary, confidential information being placed at risk.
- **Personal information.** Personally identifiable information is increasingly the subject of various international, federal, state, and local laws. While these laws each define such information differently, many of them define this information broadly to include any information that identifies or can be used to identify an individual, such as name, address, and even IP addresses, and other device identifiers. Therefore, the collection and use of personally identifiable information are increasingly handled in clauses separately from confidential information. These clauses not only include an obligation to keep personally identifiable information confidential, but also to limit its use to solely what is necessary to perform the services for the customer and to assist the customer in meeting its obligations related to requests from individuals to exercise their rights to the personally identifiable information under applicable laws.
- **Warranties.** In addition to any standard warranties relating to how the services are to be performed, freedom from viruses and other harmful code, noninfringement, and authority to enter into the agreement, the following specific warranties relating to information security should be considered:
  - A warranty requiring the vendor to comply with "best industry practices relating to information security." Such a "floating" standard will ensure

that the vendor must continually evolve its information security measures to keep pace with industry best practices. In many instances, it is appropriate to specify the industry relevant to the data (e.g., healthcare, financial services).

- Compliance with applicable consumer protection laws, such as Gramm–Leach–Bliley Act (GLB), Health Insurance Portability and Accountability Act (HIPAA), and relevant state statutes.
- If relevant, compliance with third-party standards such as the payment card industry (PCI) data security standard (available at www.pcisecuritystandards.org) or the payment application data security standard.
- Compliance with the customer's (not the vendor's policy) privacy policy in handling and using consumer information.
- A warranty against sending the customer's data and confidential information to offshore subcontractors or affiliates, unless specifically authorized to do so by the customer. The world is complex and dangerous place when it comes to data. While some countries have their own laws governing data privacy and information security, many do not. When they exist, local laws frequently conflict and do not provide the level of protection found in the United States. When data flows across international borders, many questions arise: what privacy laws apply, what happens if the data becomes the subject of a subpoena and must be produced, or do some of the countries have laws that would permit offshore suppliers to retain data after contract termination to satisfy various retention obligations imposed by law. In some cases, there are no clear answers. In others, the gray areas are very broad. Given the complexity, uncertainty, and associate risk, companies must apprise themselves of where their data will be located and make every effort to limit those locations in their contract with the vendor.
- A warranty stating that the vendor's responses to the vendor due diligence questionnaire, which should be attached as an exhibit to the contract, are true and correct.
- **General security obligations.** Consider including generalized language in the contract relating to the vendor's obligations to adopt a minimum set of security controls and to additionally take all reasonable measures to secure and defend its systems and facilities from unauthorized access or intrusion, to periodically test its systems and facilities for vulnerabilities, to immediately report all breaches or potential breaches of security to the business, to participate in joint security audits, and to cooperate with the business's regulators in reviewing the vendor's information security practices.
- **Indemnity.** In situations in which a breach of the vendor's security or inappropriate use of personally identifiable information may expose the company to potential claims by third parties (e.g., a breach of

consumer information may result in claims by the business's customers), the agreement should include an indemnity provision requiring the vendor to defend the company from those claims and to hold the company harmless from all claims, damages, and expenses incurred by the company resulting from a breach of the vendor's security or obligations regarding its processing of personally identifiable information. That is, the vendor should protect the company from lawsuits and other claims that result from the vendor's failure to adequately secure its systems or fail to live up to its obligations regarding the processing of personally identifiable information.

- **Responsibility for costs associated with security breach notification.** Breaches of security with regard to personally identifiable information may trigger obligations under a variety of state and federal laws requiring the company to send notices to affected individuals advising them of the breach. The cost of those notices may be significant, including costs related to making the required notification, providing affected individuals with identity theft monitoring and protection when appropriate, as well as costs associated with negative publicity and governmental investigation and enforcement actions. Consider inserting provisions into the vendor agreement requiring the vendor to pay for all costs incurred by the company in complying with security breach notification laws or providing such identity theft monitoring and protection. The contract should also make clear that the company has sole control over the content and timing of those notices and whether to provide such monitoring and protection.
- **Forensic assistance.** In the event of a breach, the contract should require the vendor to preserve all relevant evidence and log files and furnish that information to the company. The vendor should also provide the company with all information relating to any forensic examinations it conducts of the vendor's systems.
- **Audit rights.** The agreement should include clear rights permitting the company to audit the vendor to confirm compliance with the terms of the agreement and applicable laws and regulations. While reasonable limitations can be included regarding the number of times audits may be conducted and their timing, providers should avoid any unduly strict limitations (e.g., limiting audits to only once per year or imposing an excessive notice period before the audit can be conducted). The vendor should be required to reasonably cooperate with the audit, including providing all appropriate documentation. That cooperation should be at no cost to the businesses. Finally, the audit language should require the vendor furnish copies of all relevant third-party audit reports such as: (i) an SSAE 16 or its successor SSAE 18 "SOC 1" report for service organizations providing financial services, (ii) an SOC 2 Type I or Type II or SOC 3 reports for security and confidentiality, processing, privacy, and/or availability

controls for other processing other service organizations, such as SaaS providers (Type II is generally preferred since it covers a minimum of a six-month period instead of a point of time), or (iii) an ISO/IEC 27001 certifications for any type of organization.

- **Limitation of liability.** Most agreements have some form of "limitation of liability"—a provision designed to limit the type and extent of damages the contracting parties may be exposed to. It is not uncommon to see these provisions disclaim the vendor's liability for all consequential damages (e.g., lost profits, harm to the business' reputation) and limit all other liability to some fraction of the fees paid. These types of provisions are almost impossible to remove from most agreements, but it is possible to require the vendor to exclude from the limitations damages flowing from the vendor's breach of confidentiality and their indemnity obligation for claims the vendor, itself, causes because of its failure to adequately secure its systems. Without those exclusions, the contractual protections described above would be essentially illusory. If the vendor has no real liability for breach of confidentiality because the limitation of liability limits the damages the vendor must pay to a negligible amount, the confidentiality provision is rendered meaningless.
- **Post-contract policing.** Separate and apart from the contractual terms, the company should conduct ongoing audits, as described above, site visits, and other post-contract activities to ensure the vendor continues to comply with its information security obligations.

Information Security Requirements Exhibit: The Third Tool

The final tool in minimizing vendor information security risks is the use of an exhibit, statement of work, or other contract attachment that specifically defines the minimum security requirements relevant for a particular transaction. For example, the information security requirements exhibit may prohibit the vendor from transmitting the company's information over wireless networks (e.g., 802.11a/b/g/n/ac/ax) or over public networks without encryption or from transferring that information to removable media that could be easily misplaced or lost. The exhibit may also contain specific requirements for physical access to processing systems, the use of encryption at rest and access control technology, requirements for internal risk assessments and training of personnel, and for decommissioning hardware and storage media on which the company's information was stored to ensure the information is properly scrubbed from the hardware and media. Other specific physical and technological security measures should be identified as relevant to the particular transaction.

An example security requirements exhibit is provided at the end of this chapter.

Summary

Companies face unique risks when they entrust personal information and proprietary and confidential information to their vendors. Those risks can be minimized by employing the tools discussed in this chapter: appropriate and uniform due diligence, use of specific contractual protections relating to information security, and use, where relevant, of exhibits or other attachments to the agreement detailing unique security requirements to be imposed on the vendor.

Example Information Security Requirements Exhibit

This exhibit provides the information security procedures to be established by vendor before the effective date of this agreement and maintained throughout the term. These procedures are in addition to the requirements of the agreement and present a minimum standard only. However, it is vendor's sole obligation to (i) implement appropriate measures to secure its systems and data, including XYZ COMPANY confidential information, against internal and external threats and risks; and (ii) continuously review and revise those measures to address ongoing threats and risks. Failure to comply with the minimum standards set forth in this exhibit will constitute a material, noncurable breach of the agreement by vendor, entitling XYZ COMPANY, in addition to and cumulative of all other remedies available to it at law, in equity, or under the agreement, to immediately terminate the agreement. Unless specifically defined in this exhibit, capitalized terms shall have the meanings set forth in the agreement.

- **Security policy.** Vendor shall establish and maintain a formal, documented, mandated, company-wide information security program, including security policies, standards, and procedures (collectively "Information Security Policy"). The Information Security Policy will be communicated to all vendor personnel and contractors in a relevant, accessible, and understandable form and will be regularly reviewed and evaluated to ensure its operational effectiveness, compliance with all applicable laws and regulations, and to address new threats and risks.
- **Personnel and vendor protections.** Vendor shall screen all personnel contacting XYZ COMPANY confidential information, including customer information, for potential security risks and require all employees, contractors, and subcontractors to sign an appropriate written confidentiality/nondisclosure agreement. All agreements with third parties involving access to vendor's systems and data, including all outsourcing arrangements and maintenance and support agreements (including facilities maintenance), shall specifically address security risks, controls, and procedures for information systems. Vendor shall supply each of its personnel and contractors with appropriate, ongoing training regarding information security procedures, risks, and

threats. Vendor shall have an established set of procedures to ensure personnel and contractors promptly report actual and/or suspected breaches of security.

- **Removable media.** Except in the context of vendor's routine backups or as otherwise specifically authorized by XYZ COMPANY in writing, vendor shall institute strict physical and logical security controls to prevent transfer of customer information to any form of removable media. For purposes of this exhibit, **Removable Media** means portable or removable hard disks, floppy disks, USB memory drives, zip disks, optical disks, CDs, DVDs, digital film, memory cards (e.g., Secure Digital ("SD"), Memory Sticks ("MS"), CompactFlash ("CF"), SmartMedia ("SM"), MultiMediaCard ("MMC"), and xD-Picture Card ("xD")), magnetic tape, and all other removable data storage media.
- **Data control; media disposal and servicing.** XYZ COMPANY confidential information (i) may only be made available and accessible to those parties explicitly authorized under the agreement or otherwise expressly by XYZ COMPANY in writing; (ii) if transferred across the Internet, any wireless network (e.g., cellular, 802.11x, or similar technology), or other public or shared networks, must be protected using appropriate cryptography as designated or approved by XYZ COMPANY in writing; and (iii) if transferred using removable media (as defined above) must be sent via a bonded courier or protected using cryptography designated or approved by XYZ COMPANY in writing. The foregoing requirements shall apply to backup data stored by vendor at off-site facilities. In the event any hardware, storage media, or removable media must be disposed of or sent off-site for servicing, vendor shall ensure all XYZ COMPANY confidential information, including customer information, has been "scrubbed" from such hardware and/or media using industry best practices (e.g., DoD 5220-22-M Standard), but in no event less than the level of care set forth in NIST Special Publication 800-88, Guidelines for Media Sanitization.
- **Hardware return.** Upon termination or expiration of the agreement or at any time upon XYZ COMPANY's request, vendor will return all hardware, if any, provided by XYZ COMPANY containing XYZ COMPANY confidential information to XYZ COMPANY. The XYZ COMPANY confidential information shall not be removed or altered in any way. The hardware should be physically sealed and returned via a bonded courier or as otherwise directed by XYZ COMPANY. In the event the hardware is owned by vendor or a third-party, a notarized statement, detailing the destruction method used and the data sets involved, the date of destruction, and the company or individual who performed the destruction will be sent to a designated XYZ COMPANY security representative within fifteen (15) days of termination or expiration of the agreement or at any time upon XYZ COMPANY's request. Vendor's destruction or erasure of customer information pursuant to this section shall be in compliance with best industry practices (e.g., DoD

5220-22-M Standard), but in event less than the level of care set forth in NIST Special Publication 800-88, Guidelines for Media Sanitization.

- **Physical and environmental security.** Vendor facilities that process XYZ COMPANY confidential information will be housed in secure areas and protected by perimeter security such as barrier access controls (e.g., the use of guards and entry badges) that provide a physically secure environment from unauthorized access, damage, and interference.
- **Communications and operational management.** Vendor shall (i) monitor and manage all of its information processing facilities, including, without limitation, implementing operational procedures, change management, and incident response procedures; and (ii) deploy adequate antiviral software and adequate backup facilities to ensure essential business information can be promptly recovered in the event of a disaster or media failure; and (iii) ensure its operating procedures will be adequately documented and designed to protect information, computer media, and data from theft and unauthorized access.
- **Access control.** Vendor shall implement formal procedures to control access to its systems, services, and data, including, but not limited to, user account management procedures and the following controls:
  - Network access to both internal and external networked services shall be controlled, including, but not limited to, the use of properly configured firewalls;
  - Operating systems will be used to enforce access controls to computer resources including, but not limited to, authentication, authorization, and event logging;
  - Applications will include access control to limit user access to information and application system functions; and
  - All systems will be monitored to detect deviation from access control policies and identify suspicious activity. Vendor shall record, review, and act upon all events in accordance with incident response policies set forth below.
- Incident notification. Vendor will promptly notify (but in no event more than twenty-four (24) hours after the occurrence) the designated XYZ COMPANY security contact by telephone and subsequently via written letter of any potential or actual security attacks or incidents. The notice shall include the approximate date and time of the occurrence and a summary of the relevant facts, including a description of measures being taken to address the occurrence. A security incident includes instances in which internal personnel access systems in excess of their user rights or use the systems inappropriately. In addition, vendor will provide a monthly report of all security incidents noting the actions taken. This will be provided via a written letter to the XYZ COMPANY security representative on or before the first week of each calendar month.

Chapter 23

Distribution Agreements

CHECKLIST

License Grant

- ☐ Consider necessity and scope
  - Exclusivity
  - Territorial limitation
  - Hybrid of exclusive and territory
  - Quotas
- ☐ Reservation clause
- ☐ Noncompetition clause
- ☐ Intellectual property license
- ☐ Termination in the event of a breach

End-User License Agreement

- ☐ Which agreement will govern use of distributed product?
- ☐ Process for getting agreement to customer
- ☐ Who may accept agreement on manufacturer's behalf?
- ☐ Who owns the customer that purchases the product?

Development of the Product

- ☐ Identify parties' representatives
- ☐ Describe process for setting meetings

- ☐ Draft development plan with technical aspects of product design
- ☐ Agree ahead of time how parties will allocate expenses

Obligations of the Parties

- ☐ Distribution and sale of the product
- ☐ Training distributor's employees
- ☐ Customer support
- ☐ Marketing of product
  - – Marketing materials
  - – Marketing plans
  - – Press releases
- ☐ Assign primary marketing contact for each party
- ☐ Ensure that distributor cannot misrepresent or otherwise make false statements regarding product

Product Pricing

- ☐ Determine price of product and royalty payments
- ☐ Collecting fees
  - – Initial license
  - – Add-on services
  - – Subscriptions
- ☐ Shipping costs and taxes
- ☐ Expenses incurred in distribution
- ☐ Invoicing and collection of fees
- ☐ Distributor's periodic reports:
  - – Sales
  - – Marketing
  - – Audit payments
- ☐ Maintain records after termination of distribution agreement confidentiality

Term of Agreement

- ☐ Ensure survivorship of certain contractual clauses
  - – Warranties
  - – Indemnification
  - – Risk of loss
  - – Limitation of liability
  - – Intellectual property

Overview

Distribution agreements are used between a manufacturer of information technology (e.g., a software or hardware manufacturer) and a supplier of information technology or publisher of content to a wide base of customers (e.g., a marketer and distributor of software products). Generally speaking, a distribution arrangement provides an opportunity for both sides to make money, and a balanced relationship will be one where the manufacturer and the distributor both make monetary gains. In most cases, the manufacturer will "sell" whatever it is that it manufactures to the distributor, who will then "sell" the product for a higher price. It is often the case that the manufacturer makes the least money, but it should, nonetheless, be well compensated for the products being distributed.

These types of agreements can range from the simple (i.e., a very concise statement of the obligations of the manufacturer to provide the product, obligations and restrictions on distribution, and a statement of the business deal) to the very complex (i.e., license terms, marketing obligations, collaboration obligations if the parties are to jointly develop any product).

In all cases, it is critical that the parties clearly articulate in the distribution agreement the manufacturer's obligations with respect to providing the product to be distributed, the distributor's obligations with respect to marketing and distributing the products, any intellectual property obligations, and a clear statement of the business deal (e.g., what will the manufacturer be paid for its products and when will payment be made?).

Key Issues for Distribution Agreements

License Grant

In each distribution agreement, it is essential to consider whether a license grant is necessary and, if it is, what the scope of the license grant should be. Generally speaking, a license grant in a distribution agreement should be broad enough to cover all of the required distribution obligations of the distributor. So, for example, it should include the right for the distributor to market, promote, distribute, sell, and use the products that will be distributed.

- Consider whether the license should be exclusive so that the distributor is the only entity that can distribute the product or nonexclusive so that the manufacturer is permitted to enter into multiple distribution agreements for the same products.

- Manufacturers frequently will want to limit the distribution rights to a particular territory (e.g., a particular state, region, or country).
- A hybrid approach is often acceptable, in which the distributor has the exclusive rights of distribution within a defined territory, coupled with nonexclusive rights outside of that territory. Be careful—this will restrict the manufacturer's ability to enter into exclusive relationships where a particular distributor is granted nonexclusive distribution rights.
- Quotas are often used in order to ensure at least a specified number of units distributed within the specified period of time. Generally, if the determined quota isn't met, the parties will meet to determine what factors contributed to the failure to meet the quota, whether those factors can be reasonably addressed and corrected, and how much time the distributor will have to reach the quota as set by the parties. Failure to meet the determined quotas can also result in a termination right for the manufacturer.
- Consider whether a reservation clause is appropriate. These types of clauses make it clear that the manufacturer has the ability to directly or indirectly market, distribute, demonstrate, sell, resell, license, maintain, support, and otherwise commercially exploit its products in the defined territory and through any other sales channels, resellers, and distributors, in its discretion.
- While distributors often seek exclusive distribution rights, manufacturers may also want to ensure that the distributor's hands are tied with respect to entering into distribution arrangements for competitive products. A noncompetition clause can protect a manufacturer from working with distributors who market and distribute competing products, thereby cannibalizing potential sales of the manufacturer's products. This restriction can, and often does, also extend to limitations on the rights of the distributor to work with competing companies to develop products that can or may compete with the manufacturer's products.
- Licenses to trademarks and other intellectual property may be necessary in distribution agreements that contain a marketing component. These licenses are commonly limited to the other party's right to use, reproduce, display, and distribute the marks of the other party in connection with and as necessary to market and distribute the product to customers in accordance with the distribution agreement. The party granting the license should be sure to include a requirement that the other party comply with the licensor's acceptable use policies, intellectual property policies, and any other advertising and marketing policies in the use of the licensor's marks and other intellectual property.
- Manufacturers will commonly want to terminate the distribution agreement in the event that the distributor breaches certain obligations under the license.

End-User License Agreement

The parties must collaborate to determine what agreement will ultimately govern the use of the distributed product. Distributors will frequently have little or no say in this regard, as manufacturers should be able to ensure that their products are used in accordance with their terms and conditions (e.g., in the software context, the manufacturer's end-user license agreement). The distribution agreement should contain a process for how the applicable agreement will get to the customer (i.e., will the distributor or the manufacturer initiate that process) and who may accept the agreement on the manufacturer's behalf. In some cases, the distribution agreement will require that the distributor ensure that each customer is bound by terms and conditions at least as protective of the manufacturer as those contained in the manufacturer's standard terms and conditions, which would typically be attached to the distribution agreement as a schedule.

The overall structure of the contract is designed to ensure neither the distributor nor the manufacturer is placed at risk. In general, each will have a direct contract with the customer. Each is able to draft its own desired terms.

One of the most important drafting points in a distribution is to ensure the contract does not create the possibility of an "end run" around the end-user agreement. An easy example of this would be a request by a distributor for an indemnity from the manufacturer for any claim that arises from the product being distributed. Other than the request for intellectual property infringement indemnification, this type of provision creates a substantial hole in the protection of the end-user agreement. The end user's remedies against the manufacturer are strictly limited by the end-user agreement. So, instead of suing the manufacturer directly, the end user sues the distributor, and the distributor then seeks indemnification from the manufacturer. This, essentially, renders the protections in the end-user agreement illusory. The only way to protect against this possibility is to refuse this type of omnibus indemnity and insist the distributor have its own, well-drafted agreement in place with the end user. Such an agreement would generally make clear the distributor is not responsible for claims relating to the product being distributed and that the end user's sole remedies are those provided in the manufacturer's own end-user agreement. A key consideration in any distribution agreement is who ultimately "owns" the customer who purchases the product. The manufacturer will commonly want control over the end user and the data generated from sales of its products. However, this information is also extremely valuable to the distributor. Care should be taken to ensure that the rights with respect to ownership of the customer are precisely articulated in the distribution agreement.

Development of the Product

Where the manufacturer and distributor collaborate to develop the product that will ultimately be distributed, the distribution agreement will contain a detailed process and precise obligations of both parties with respect to the development.

- Identify the representatives of each of the parties responsible for oversight of the development process.
- Describe the process for scheduling and conducting regular meetings and the location (physical or virtual) of those meetings.
- Draft a development plan, attached as a schedule to the distribution agreement, that clearly sets forth all of the technical aspects of the design of the product.
- Ensure that the parties have agreed ahead of time with respect to the allocation of expenses during the development process (i.e., will the parties be responsible for their own expenses, or will some other allocation be agreed upon?).

End-User Data

Data of customers of the distributed product can be extremely valuable to both parties to the distribution agreement. Each distribution agreement should clearly articulate whether that data is owned by one or the other party to the distribution agreement or is shared by both parties.

Obligations of the Parties

The specific obligations of the parties to a development agreement will depend entirely on the scope of the business deal. In many development agreements, it is common to include the following obligations:

- The distributor is generally responsible for the distribution and sale of the product. Where the product is a subscription-based application, this will include a responsibility to distribute and sell the initial subscription, coupled with marketing and sale obligations with respect to renewal subscriptions.
- The manufacturer should be responsible for training all of the distributor's employees with respect to the marketing, sale, and distribution of the manufacturer's product. In the case where a product developed mutually by the manufacturer and distributor is being distributed, the distributor will be responsible for any training associated with any portion of the product that it developed.
- Obligations of customer support are handled on a case-by-case basis. In some cases, the initial line of support for the distributed product will be the responsibility of the distributor, with second-line support handled by the

manufacturer. In other cases, the manufacturer will want to maintain sole control over the support of its products.

- Marketing of the product is commonly an obligation of the distributor, and the distribution agreement should precisely articulate the specific efforts that the distributor will be required to make with respect to the marketing of the product. It is commonly appropriate for the manufacturer to assist the distributor with marketing the product, including development of the marketing materials and a marketing plan. Both parties can agree to distribute press releases (approved by both parties) announcing the distribution relationship.
- Each party may want to assign a primary contact to work with the other party on marketing, promotion, and sales efforts.
- The manufacturer should ensure that, with respect to the marketing of the product, the distributor cannot:
  - Make any reference or claim about the manufacturer or the product that has not been preapproved by the manufacturer.
  - Use any deceptive, misleading, or illegal practices in promoting or selling the product.
  - Fail to inform the manufacturer as to problems and complaints encountered with the product.
  - Make any representations or warranties with respect to the product.
  - Use only marketing tools approved by the manufacturer.
  - Assign or sublicense any of its rights or obligations under the agreement.

Product Pricing

Each distribution agreement will clearly articulate who is responsible for determining the price of the product and the royalty payment to be made to the manufacturer for each product sold. The distribution agreement will also articulate who is responsible for collecting fees (usually the distributor). In the software context, this will include the price of the initial license as well as the price for any add-on services such as support and maintenance. In the subscription-based model, this will include the price of the initial subscription as well as each renewal subscription. The parties should also determine in the contract who is responsible for shipping costs and taxes (usually the distributor) as well as the allocation of responsibility for expenses incurred by the parties in the distribution effort (usually each party is responsible for any expenses it incurs). It is also common for the distributor to be responsible for invoicing and collection of fees for any products distributed.

- The distributor's obligations commonly include a requirement to report on all sales of products periodically (e.g., on a monthly basis). The distributor should also be required to report on marketing efforts undertaken and the names and addresses and other contact information for prospects. Audit rights are also common in distribution transactions where royalty payments are to be made. If it is discovered that the audit payments were inaccurate during the period reviewed as part of the audit, the distributor should be required to immediately pay the discrepancy (or, in the event of an overpayment, the manufacturer should pay the overage). It is also common in distribution agreements to shift the burden of payment for the audit to the distributor if the underpayment amount reaches a predetermined percentage of total fees paid (e.g., more than 10% or 15%).

- It is important that both parties maintain records for a lengthy period of time following any termination or expiration of the distribution agreement to ensure that accurate audits can be conducted.

Additional Considerations

- **Confidentiality.** While it may certainly be the case that little or no confidential information exchanges hands as part of a distribution relationship, if confidential information will be exchanged, the distribution agreement should include a standard confidentiality clause (under most circumstances mutual) protecting the confidential information that is shared. Include obligations that each party maintain the confidential information of the other party, use the other party's confidential information only as required to perform its obligations under the agreement, and return the other party's confidential information upon request and upon any termination or expiration of the agreement.
- **Term.** The parties to the distribution agreement will want to agree upon an initial term of the relationship and renewal rights. It is common to include an automatic termination in the event that the distributor does not reach a specified number of distributed products within an agreed period of time (e.g., six months). Additional termination rights commonly include:
  - Failure of a party to comply with the terms of the agreement (after, perhaps, notice and a reasonable cure period).
  - The other party's bankruptcy.
  - The mutual agreement of the parties to part ways.
  - If either party is acquired or merges with another entity.

Upon any termination, it is important to ensure that the existing licensees or purchasers revert back to the manufacturer for continued support and future sales of the software or goods that were distributed. Where software has been incorporated into the distributor's software, it is common to permit the distributor to continue distributing the combined product after termination for a specified period of time (e.g., six months) or until all of the combined products have been licensed. In this case, it is critical that certain terms of the agreement (e.g., payment) survive termination of the agreement. Further, after termination or expiration of the agreement, the distributor's rights are limited to distributing combined products created prior to the termination or expiration. The distributor would commonly be restricted from creating new combined products after termination or expiration.

- **Warranties.** Generally speaking, the manufacturer will not make any warranties with respect to its product in the distribution agreement. Any warranties related to the product to be distributed will be found in the end-user agreement, not the distribution agreement. Warranties may be added relating to each party's authority to enter into the agreement and compliance with all laws and regulations applicable to each party's obligations under the agreement.
- **Indemnification.** Indemnification by both parties of the other party is common in distribution agreements. The manufacturer will commonly indemnify the distributor for third-party claims that any marks licensed to the distributor infringe or misappropriate the third party's intellectual property rights. Common exclusions to the requirement of the manufacturer to indemnify the distributor include modifications to the manufacturer's marks, combination of the manufacturer's marks with other products or services, and the failure of the distributor to use updates to the manufacturer's marks. The manufacturer's obligations to indemnify should also include a remedy that if the mark is subject to a claim of infringement or may be subject to such a claim, then the manufacturer can procure for the distributor the right to continue to use the mark or replace the mark with a noninfringing mark. If neither of these options is available to the manufacturer, then the manufacturer is commonly afforded the right to terminate the agreement.

The distributor commonly has broader indemnification obligations that include indemnification for third-party claims to the extent arising out of a combination of the manufacturer's product with other products or any alteration of the manufacturer's products.

- **Risk of Loss.** The distributor should take responsibility for the risk of loss of the product after it is delivered to the distributor from the manufacturer. It is common in these types of agreements for the manufacturer to disclaim all liability for loss or damage caused by the distributor's failure to perform the distributor's obligations to protect the products from damage.
- **Limitation of liability.** Distribution agreements will generally include a limitation of liability that disclaims certain types of damages and limits recovery on other types of damages. While the presence of such terms is generally not objectionable, care should be taken to ensure that the scope of the limits is appropriate for the transaction.
  - The limitation of liability should be mutual—it should protect both parties.

- Most limitations of liability are drafted to exclude consequential damages and to cap recovery for direct damages at some specified amount or a multiple of fees that are paid under the agreement. The preference is to ensure that the cap is tied to all of the fees paid under the agreement during the life of the agreement. Since fewer fees will likely be paid early in the relationship, it is necessary to consider a minimum level of liability until enough fees have been paid to make the cap meaningful. Such an initial cap is commonly a percentage (e.g., 50% or 75%) of the anticipated fees to be paid during the first year of the arrangement.
- Ensure that damages arising out of certain breaches and expenses, to be paid pursuant to certain terms in the agreement, are excluded from the limitations and exclusions of liability under the agreement. For example, it is common to exclude from caps and exclusions damages arising out of a party's breach of the obligations with respect to confidentiality under the agreement. Another common exclusion is damages arising out of and expenses to be paid pursuant to a party's obligation to indemnify the other party under the agreement. Finally, it is common to draft the contract so that damages arising out of a party's breach of its obligations to keep the other party's information and data secure are carved out (i.e., not included) in the damage caps and exclusion of certain types of damages.

■ **Intellectual property.** Most distributors will want the manufacturer to acknowledge and warrant that that there are no suits, claims, or other proceedings against the manufacturer that would prevent the distribution of the software or goods being distributed. Further, distributors commonly require a warranty that the manufacturer has not violated any intellectual property or proprietary rights of a third party in developing the software or goods to be distributed by the distributor. The manufacturer should clearly retain all ownership interests in the software or goods being distributed and ensure that ownership of the goods and software is not, under any circumstances, transferred to the distributor.

Summary

Distribution agreements can take many forms. They can be for the distribution of software or goods and can include obligations with respect to collaboration, development, and marketing. The most important elements of these types of agreements are ensuring that the distributor's rights are broad enough to enable the distributor to distribute the software or goods as required by the contract, including protections of the manufacturer with respect to intellectual property ownership of the manufacturer's software or goods, ensuring that the manufacturer has access to the information of customers who license or purchase the software or goods that are being distributed, and setting up a fees and payment structure that promotes performance of the parties.

Chapter 24

Data Agreements

CHECKLIST

Key Contractual Protections

- ☐ Include basic contractual protections common to all technology agreements
  - Confidentiality
  - Limitation of liability
  - Termination rights
- ☐ Ensure the scope of the license is broad enough to include all intended uses, both those existing at the time the contract is signed and reasonably anticipated in the future
- ☐ Pre-negotiate fees for increasing the scope of the license
- ☐ Avoid overreaching audit rights
  - Limit frequency of audits
  - Limit the type of records that can be reviewed
  - Limit the duration of the audit
  - Include protections relating to third-party audits (e.g., require auditors to sign an NDA and to be mutually agreed upon by the parties)
  - Limit costs recoverable for third-party auditors
  - Reject requests from the vendor to recover internal personnel costs
- ☐ Warranties
  - Rights to grant the license
  - The vendor has no knowledge of infringement claims
  - Reasonable efforts to ensure timeliness and accuracy of data
  - Reasonable efforts to notify the customer of known errors in the data
- ☐ Include an indemnification against infringement claims based on the customer's licensed use of the data

- ☐ Negotiate broad termination rights, including termination for convenience, wherever possible
- ☐ The contract should specify the manner and format of delivery for the data
- ☐ The customer should ensure it has the unilateral right to renew the contract for at least a few years
- ☐ The agreement should include price protection for the initial term and first few renewal terms

Overview

Data or data feed agreements are a special type of license agreement that involves access to a collection of data. That data is typically harvested by the vendor from various public sources. The data can relate to mapping data for real estate–related transactions, market data relating to the financial markets, data gleaned from public records (e.g., real estate transactions, business licenses, marriage licenses), or aggregated data harvested from user interactions with websites or search engine requests. The common thread is that the data is generally not created by the vendor, but merely collected or harvested from various public sources and then licensed as a package. In many instances, the vendor may have invested extraordinary time and money in creating the database.

Data agreements present unusual contracting issues. First, the vendors generally insist on using their own form agreements. Second, they also strongly resist any substantive revisions to those agreements. Finally, the data being licensed is normally provided on essentially an as-is and as-available basis, without warranties of any kind.

Key Contractual Protections

Given the foregoing challenges, this chapter discusses the revisions that can generally be negotiated with these types of vendors:

- **Basic contract protections.** Certain protections should be considered regardless of type of technology contract. These provisions should be included in data agreements (e.g., confidentiality, limitation of liability).
- **License scope.** One of the most critical elements of a data agreement is the scope of the license being granted. That is, the specific language in the contract that says what the customer may and may not do with the data. This is the area where most errors are made. Customers fail to really think through all the purposes they may have for the data. Customers must carefully consider this issue, including potential future uses, and ensure all uses are clearly described and included in the agreement.

- **Fees for increased use.** Most data is licensed depending on the breadth of its use. Increased use typically requires the payment of additional fees. It is important to pre-negotiate fees for potential areas where use may increase or change over time and include them in the contract (e.g., increased users accessing the data, increased areas where the data may be used). Failing to do so may result in the customer having little or no negotiating leverage when it comes time to discuss fees for increased use with the vendor.
- **Limitations on audit rights.** Most data agreements include extensive audit rights permitting the vendor to enter the customer's premises and access its systems and records. These audits can be very invasive and disruptive. In addition, if the customer, itself, has highly sensitive data (e.g., personally identifiable data), granting access to this type of broad access may create a security risk. These types of audit rights should be eliminated wherever possible and replaced with a more limited right to inspect records created in the ordinary conduct of the customer's business. The following is an example of a more appropriate audit right:

Compliance with license. On vendor's written request, no more than once during any twelve (12)-month period of the term of this agreement, vendor or a mutually agreed upon nationally recognized auditor may review customer's generally available records regarding its compliance with the agreement. For the avoidance of doubt, customer shall be under no obligation to create or compile information that is not readily available in the ordinary course of its business. Customer shall make the records available in a conference room at its facilities at a mutually agreed upon date and time. The foregoing review right shall not afford vendor any other access to customer's facilities, systems, and records. The review shall not unreasonably interfere with customer's operations. The review shall be completed within two (2) business days. Prior to being granted access to customer records or facilities, vendor's third-party auditor shall execute customer's required nondisclosure agreement. Vendor and any third-party auditor shall comply with customer's then-current standard access and security policies while present at customer facilities. All information obtained during the course of an audit shall be deemed customer confidential information.

- Be wary of audit clauses that shift the cost of the audit to the customer if noncompliance is identified (e.g., the customer has underpaid license fees). Those costs can be substantial, sometimes running from tens of thousands to hundreds of thousands of dollars. If this language cannot be avoided, include a limitation that the costs of the audit cannot exceed, for example, 25% of the amount of the underpayment.
  - Limit the costs recoverable to the "reasonable" costs of the third-party auditor. Reject requests to also compensate the vendor for its internal personnel time. Such requests are not consistent with industry practice.

  - Also, most licensor form audit clauses have the cost of the audit shift if there is a 5% or more difference in fees or usage. For licensees, that percentage should be negotiated to 10% or 15%.
  - Finally, include a statement that no auditor can be engaged on a contingent fee basis. That is, auditors that only get paid if they actually find a noncompliance. Such audit engagements can easily get out of hand as the auditor conducts a protracted audit, knowing it won't be paid if does not uncover a noncompliance.
- **Warranties.** In most cases, vendors will not offer much in the way of substantive warranties. That said, the following warranties can usually be negotiated (revised to reflect the defined terms in the agreement):
  - Vendor has all rights necessary, including rights of third parties, to grant the license provided under this agreement.
  - To the best of vendor's knowledge as of the effective date, customer's licensed use of the data shall not infringe the intellectual property rights of any third party.
    - Vendor shall use commercially reasonable efforts to ensure the accuracy and timeliness of the data licensed.
    - Alternate language to the foregoing warranty: Vendor shall use such efforts as are common to other licensors of markets of this kind to ensure the accuracy and timeliness of the data licensed.
  - Vendor shall use reasonable efforts to promptly notify customer of any errors in the licensed data of which vendor becomes aware.
- **Indemnity.** The vendor should be required to provide a basic indemnity in the event the customer's "licensed use of the data" infringes the intellectual property rights of any third party. Even though the underlying data may have been obtained from third parties and the vendor may be reticent to offer an indemnity of this kind, it is common in the industry to be able to negotiate this protection in most data agreements.
- **Termination rights.** Given the fact that most data agreements have little in the way of contractual protections or performance requirements, whenever possible, termination for convenience rights should be included: "Customer may terminate this Agreement at any time, without cause or penalty, on thirty (30) days prior written notice to Vendor."
  - The foregoing termination right is frequently difficult to negotiate. In such cases, consider offering to toll the termination right for the first six months of the contract or to provide for a longer notice period prior to termination: "At any time after the initial six (6) months of the term, Customer may terminate this Agreement at any time, without cause or penalty, on sixty (60) days prior written notice to Vendor."
  - The customer should also have the right to terminate without penalty if the data being licensed becomes the subject of an intellectual property

  infringement claim or if the vendor repeatedly fails to deliver the data as required under the agreement.
  - **Data format.** Ensure the data will be provided in a format that will be usable to the customer. That is, the contract should specifically identify the means of providing the data and the electronic format of the data.
  - **Renewal rights.** The customer should negotiate the ability to unilaterally renew the agreement for a defined number of years. For example, the customer might negotiate the ability to unilaterally renew for up to three one-year terms following the initial term. After that, the contract may automatically renew for additional one-year terms, but either party can elect not to renew by, for example, providing notice to the other party at least ninety days prior to the next renewal data.
- **Price protection.** Negotiate price protection during the term of the agreement as in the following language:

Fees during renewal terms. Vendor's fees hereunder shall be fixed during the initial term. Thereafter, vendor may increase such fees for a renewal term by providing notice to customer at least sixty (60) days prior to the commencement of such term. Any increase shall not exceed the lesser of: (i) four percent (4%) of the fees charged during the preceding term; or (ii) the amount equal to the percentage change in the consumer price index ("CPI") during the preceding calendar year. For purposes of the foregoing, the CPI shall be the index compiled by the United States Department of Labor's Bureau of Labor Statistics, Consumer Price Index for all Urban Consumers ("CIP-U") having a base of 100 in 1982–1984, using that portion of the index that appears under the caption "Other Goods and Services." (For reference, the CPI is currently posted on the website of the Bureau of Labor Statistics at www.bls.gov.) The percentage change in the CPI shall be calculated by comparing the annual CPI figures and expressing the increase in the CPI as a percentage. If the CPI is no longer published, or the CPI currently published is materially changed, vendor and customer will negotiate, in good faith, revisions to this section to reflect and account for such changes in the CPI.

Summary

Data agreements present unique issues. Unlike most other types of license agreements, the vendor may not be the creator of the content being licensed. Rather, that content may be harvested from hundreds or even thousands of other databases and public records. Vendors may be very reticent to provide many of the protections found in other licensing engagements. In these cases, the focus should be on basic protections: intellectual property indemnity, termination rights, and basic assurances that the data will be current and free of known errors.

Chapter 25

Website Development Agreements

CHECKLIST

Criteria to Consider When Selecting a Website Developer

- ☐ Developer's experience and qualifications
- ☐ Identify parties and their relationship
- ☐ Extent of allowable subcontracting
- ☐ Whether to include both designing and hosting
- ☐ Implications of co-branding or joint development

Basic Objectives

- ☐ Specifications for website
  - – Statements of work
  - – Technology and equipment to be supplied
- ☐ Change management process
- ☐ List competing websites and extent to which your website is based on them
- ☐ Required functions
- ☐ Maintenance and updating requirements
- ☐ Define developer rights of access

Intellectual Property Ownership

- ☐ Obtain proper licenses and assignments

Software Requirements

- ☐ Ensure "open use" software
- ☐ Determine rights to software and whether to sublicense or separately license
- ☐ Acquire disclosures from developer
- ☐ Identify who owns the software
- ☐ Address disabling devices

Schedules and Timetables

- ☐ Start date
- ☐ Anticipated termination date
- ☐ Intermediate "checkpoints"
- ☐ Process for modifying schedule
- ☐ Consequences if "checkpoint" not met:
  - Extension
  - Monetary penalties
  - Termination
  - Delay
  - Acceleration

Term and Termination

- ☐ Initial term
- ☐ Maintenance, hosting, and colocation services
  - Requirements for developer:
  - Return company property
  - Transfer software to company
  - Turn over documents
  - Confidentiality assurances
  - Receipts for reimbursements
- ☐ Final statement from developer and company that work was completed
- ☐ Termination without consent for material breach of contract

Other Provisions

- ☐ Fees, charges, and expenses
- ☐ Project management
- ☐ Acceptance testing
- ☐ Warranties
- ☐ Identifications
- ☐ Content of website
- ☐ Linking issues

- ☐ Insurance
- ☐ Reports, records, and audits
- ☐ Training, education, and troubleshooting
- ☐ Disputes
- ☐ Trademarks and copyright
- ☐ Privacy
- ☐ Terms of use

Overview

Website development agreements describe the programming, services, and other requirements associated with a company's website development project. Websites come in many different varieties, ranging from the plain to the glitzy and from the simple to the complex. After a company has made some basic decisions concerning the type of site it wants—or the type of site it wants its current site to become—a website developer should be selected who has the experience and capability in that type of site. To some extent, the criteria and desires of the company in this process should be reflected in the website development agreement.

Initial Issues to Think About

- Companies should evaluate a potential website developer's experience and qualifications. It is frequently appropriate to include in the website development agreement a recitation of the developer's experience, sometimes as a substitute for a warranty from the website developer concerning the developer's ability and qualifications to do the job.
- It is important for both sides to know who the parties to the agreement are and what their relationship is. On the developer side, it is important to know whether the "contractor" is an entity that will actually do the work through individuals and whether those individuals are going to be independent contractors or employees. On the company side, the issue is sometimes related to "affiliates" (who would need to be defined) and subsidiaries. In addition, consider whether the parties are entering into an exclusive relationship, or if either party can work with a competitor of the other party during and after the term of the agreement.
- The website development agreement should address the extent to which subcontracting of the work will be allowed or prohibited. If subcontracting is permitted, the company should have approval rights over which subcontractors are used, or the parties should agree ahead of time to a list of approved subcontractors. In each case, the company should ensure it conducts proper diligence with respect to, at a minimum, experience and qualifications, of any

subcontractor who will be performing services associated with the development of the website for the developer.

- Consider whether the developer is merely designing the website or if it will also host the site. If the developer is hosting the website, the agreement should include essential hosting requirements.
- If the website is co-branded or jointly developed, the agreement will commonly describe who will own any end-user information collected from operation of the website. In addition, it is important to describe how this information will be used and what happens to this information upon termination of the agreement.

What are the Basic Objectives of the Website and the Development Agreement?

This is perhaps the heart of the website development agreement. It should describe and define exactly what the parties intend to develop. The language of the agreement should be as specific and detailed as possible. The following factors are essential considerations for any website development agreement:

- The specifications for the website are commonly attached to the website development agreement as an exhibit. This may take the form of a statement of work (SOW), which is sometimes a single statement and is sometimes a series of statements that evolves as additional features or capabilities are added. If your company anticipates a series of SOWs, ensure that the agreement includes a process for developing and approving those SOWs and for including them as exhibits to the agreement. It is important to describe in the agreement or in an exhibit to the agreement what technology, and equipment, each party must supply to the other in order for the website to meet the specifications.
- Website development agreements commonly include a detailed process for changing or amending the specifications and the SOWs. This is often done in the form of change orders that must meet certain requirements, such as being in writing and issued only by authorized persons. A change management process is essential if changes to specifications and SOWs are anticipated.
- The parties must understand ahead of time the extent to which the new website is to be based on another website. For example, the website may need to have at least the same capabilities as the websites of major competitors. If that is the case, the competitors' websites would be listed in the contract.
- Each website development agreement will specifically state the functions that are required. These are often listed in the specifications, SOWs, or other attachments to the agreement. For example, in many website development transactions, there are certain search, indexing, linking, expansion, and maintenance requirements. In addition, the specifications and SOWs will also

include specific required features and a description of those features. These may include requirements with respect to tool bars, buttons, online forms, graphics, and music capabilities (a separate legal issue such as music copyrights and the permissions necessary to use music on a website is a complex legal issue). Another feature the company may want to consider specifying is a maximum download time, or some stated rate of speed; the graphics and other features of the website may have to be matched against that requirement.

- Maintenance and updating requirements are frequently included in website development agreements, either in the terms and conditions or in an attached maintenance and support agreement. The maintenance and updating requirements frequently address the following topics:
  - The extent to which periodic updates to the website are to be provided by the developer.
  - Requirements for when updates can occur (e.g., once a month, limited to two hours unless additional compensation is agreed to).
  - A requirement that the developer provide training and a manual to company personnel who will do most of the updating and routine maintenance of the site content.
- Certain programs, facilities, and rights of access will be granted by the company to the developer. In order for the developer to accomplish the desired objectives, the company will probably have to grant the developer access to various facilities and programs of the company. It is generally recommended that the company be as specific and detailed about the access or other items that it will provide to the developer as it is for the things that it expects from the developer. Problems in this area often result in finger pointing (i.e., the developer claiming that it did not meet the timetables because the company did not allow the developer sufficient access to company computers). The times and procedures for responses and participation by the company should be as detailed and specific as those for the developer.
- If the existing website is to continue in use, has it been audited to be sure everything remains appropriate to the new site, including links, copyright issues, trademark issues, terms of use, privacy, and all other aspects of the website? The company should consider the extent to which it will warrant to the developer that anything that the company authorizes the developer to use from the company's existing website is free of legal and other problems.

Intellectual Property Ownership

The website agreement should clearly describe which party owns the website. If the website is not a "and work made for hire," the company must obtain the proper licenses and assignments from the developer. These requirements will include a description of which party owns any developed intellectual property, what licenses are afforded each party for use of any newly developed intellectual property, and which party has the responsibility for registering newly developed intellectual property. Finally, if the website is not owned by the company, the agreement must describe what happens to the website and its content upon termination of the agreement, and the company and the developer should be sure to retain ownership of any preexisting intellectual property.

Beware of merely characterizing development work as "work for hire." That concept applies only in the context of copyright law, not patents or trade secrets. It provides only a very superficial form of ownership. In general, ownership language should be structured to ensure the company obtains all intellectual property rights, not merely copyright.

For a further discussion of intellectual property issues as they relate to websites, see Chapter 18.

Software Requirements

The website development agreement will include a description of the software that the developer will use to develop the website. This presents a variety of issues. The key here is to focus on the future. What could happen that would cause the company problems in continued use, maintenance, development, and expansion of its website? Remember, the company will be using the website well into the future, and certain protections and assurances with respect to the software used to develop the website must be included in the website development agreement to ensure issues do not later arise.

- The company will likely want to ensure that no "open-source" software will be used or, if it is, that the implications of that use have been thoroughly analyzed, including reviewing all license agreements applicable to any open-source software that is used to develop the website.
- If the developer is going to use software for which it only has a license rather than ownership, the company should have the opportunity to examine the license or obtain representations and warranties from the developer that the developer has the right to use that software.
  - In some circumstances, it is beneficial for the company to be granted a sublicense to certain software that the developer uses in the development of the website.
  - After development work on the website is complete and the website is accepted by the company, the company must have the ability to continue to use this software or do whatever may be necessary or appropriate to develop or expand its website. The website development agreement should contain terms ensuring that the company has this ability.

- Consider the extent to which continued use of the software will involve going back to the same developer so that the developer becomes "locked into" the operation of the company website. If this is not the desired result, the company should consider whether separately licensing the software is preferable or whether other alternative approaches are appropriate in order to ensure maximum flexibility to use other developers, or to develop in house, in the future.
- There are certain disclosures (including possibly the source code to the website) that may be appropriate for the company to have at the beginning of the project to assure that, should the developer go out of business, abandon the work, or otherwise become incapable of performing the work, the company can continue the work either in-house or through another developer, without starting over from the beginning.
- Ownership of software developed during the development process is a significant issue to consider. The website development agreement should clearly specify who owns software that is developed during the process.
- It is often the case that developers insert disabling devices into the software. Care should be taken to address this issue and, if this is prohibited, the website development agreement should make that clear. This might be addressed in the representations or warranties section of the agreement.
- Consider whether it is appropriate to put the software used to develop the website into a source code escrow account.

Schedules and Timetables

As with any information technology services agreement, schedules and timetables, often described in a detailed project plan, are of paramount importance. Care should be taken to ensure that your website development agreement establishes appropriate schedule and timetables such as the following:

- What is the start date? (In conjunction with this, it is helpful to list all of the tasks and requirements that may be necessary to begin the project.)
- What is the anticipated termination date? (What is the date on which all of the functions and capabilities specified in the agreement should be achieved?)
- What intermediate "checkpoints" or "milestones" along the way might be appropriate?
- What is the process for modifying the schedules and timetables? (Often "mutual consent" is used, and the procedures for that should be described.)
- What are the ramifications if a milestone is not met?

- Will there be an extension? If so, for how long?
- Are there any monetary penalties for being late or incentives for finishing early or reaching milestones early?
- At what point can the company terminate the agreement if it feels that delays are unreasonable, and that such delays are entirely the fault of the developer?
- Is the company allowed to delay the project? If so, what are the procedures and implications of that?
- Is the company allowed to accelerate the project? If so, what are the procedures and implications of that?

Term and Termination

The initial term of the agreement is likely to be the period in which it is thought the job can be completed. However, many website development agreements also include ongoing work, such as maintenance, hosting, and colocation services. These other matters may be contained in separate agreements, but they may also be worked into the development agreement in such a way that it becomes appropriate to specify exactly when the development phase of the project ends. Further, it may be necessary to extend the term of the website development agreement for so long as there is an outstanding and uncompleted SOW. In transactions with multiple SOWs, it may be necessary to incorporate a process whereby the term of the agreement will only end after completion of all of the services to be provided under SOWs. Also, on termination, consider the following requirements:

- The developer should return anything that it was provided by the company to do the work.
- The developer should be required to sign any documents necessary to transfer to the company any of the software that was created.
- The developer must turn over any other documents that may relate to the website or the work that was done. Examples include any consents the developer obtained from individuals whose names or pictures may have been used, licenses or consents from the owners of any trademarks that may have been included, documents authorizing the use of any music, and documents assigning copyrights if some of the work was done by independent contractors.
- On termination, there may be some final tasks to do relating to confidentiality, such as exit interviews and reaffirmations of the confidentiality provisions of the agreement by all of the developer's employees or contractors who obtained confidential information.
- If the developer paid any bills relating to the job that were reimbursed by the company, the receipts for those, if not already provided, might be included on the termination checklist.

- There might be a final acknowledgment that the work has been completed and that the developer has been paid all to which it is entitled. Note that the developer may want the corollary to this—a final statement by the company that it is satisfied with and has accepted the work that was done.

A termination of the website development agreement without the consent of the other party is potentially troublesome; therefore, anything that gives rise to such a right should be described out in detail. Often this is done by saying that either party may terminate the agreement for a breach of any material term by the other if the breach is not cured within a certain number of days. This provision can sometimes be fairly lengthy and detail the exact procedures and notices that must be given. Obvious material breaches would include the failure of the company to make the progress payments called for by the agreement, the failure of the developer to reach the milestones called for by the agreement, or the failure to create a website that is acceptable to the company under the acceptance provisions of the agreement.

As with many other information technology agreements, certain provisions of the website development agreement should be stated to survive any termination or expiration of the agreement. Consider including in the survival clause obligations of the parties with respect to confidentiality, noncompetition, dispute resolution, certain warranties and representations, limitations of liability, indemnities, and ownership of intellectual property.

Fees and Charges

The fees, charges, and anticipated expenses associated with the services should be clearly stated in the body of the agreement as well as in a fees exhibit. The terms to be contained in the body of the agreement are typically terms with respect to the following:

- How the fees and charges are to be calculated and when they are due. Is there any grace period?
- Interest or other penalties for late payments.
- Process for expense reimbursement and requirements for documentation of expenses (e.g., is the developer be required to abide by the company's standard policies on expense reimbursement?).
- An "all fees" clause that states specifically that all of the fees are stated in the agreement and there are no other fees to be paid by the company to the developer, except for those stated in the fees exhibit.

A separate fees exhibit is commonly included. Such an exhibit typically contains the actual fees to be paid by the company to the developer and the payment schedule. It is important to be as precise as possible with respect to the fees. If additional fees are anticipated, care should be taken to negotiate the rates ahead of time and include those in the fees exhibit to the website development agreement.

Project Management

Keep in mind that the development of a website is a project that should be managed carefully in accordance with terms agreed to by the parties in the website development agreement and corresponding SOWs. It is common to include a detailed project plan with respect to this type of development work including when the work will begin, who will manage it, what the milestones are for delivery of the services, and when the work will be completed. Each party should assign the appropriate number of project managers to oversee the work to ensure that the services are performed in accordance with the agreed-upon project plan.

- Generally, at a minimum, each party should appoint a project manager who understands that the success or failure of the project will play some role in that person's compensation and advancement within the company.
- In more elaborate projects, there may be various committees established to have a role in the development. An example might be a "technical committee" that would meet periodically to review any technical issues or difficulties with the project.
- The agreement should contain details with respect to what records the project managers keep and what reports they should make others within their organizations, including details with respect to the frequency of reports and to whom the reports are made.
- Project managers are frequently provided authority over certain aspects of the relationship. This can include the ability to extend time frames, increase or decrease costs, change specifications, waive failures to meet certain criteria, and establish new requirements, among other things. Each of these should be determined ahead of time and described, in detail, in the website development agreement.

Acceptance Testing

As with most information technology services agreements, the company should be granted the opportunity to test the services and deliverables. The process for acceptance testing is usually described in the terms and conditions to the agreement, while the acceptance testing criteria, specific to the services and deliverables, are more commonly found in the specific SOW. When drafting the acceptance testing portion of the website development agreement, the following factors should be considered:

- Is this an "all or none" situation where the work is simply presented to the company, tested against the specifications, and then either accepted or rejected? Or are there various elements along the way that can be meaningfully tested?
- Are there specific approvals that the company should grant along the way—falling short of an "acceptance test" but, nevertheless, heading off potential problems? (e.g., an evaluation of the proper use of the company's trademarks or other intellectual property). Also, consider whether periodic approvals with respect to the way items are organized on a computer screen are required (i.e., while some website development work is technical, some is partially "artistic" or very subjective, and while the website might meet all of the technical specifications, it does not look as the company intended). A potential issue is when the website developer assumes more "computer savvy" on the part of the company's intended customers and website visitors than the company feels is appropriate, and a more "user-friendly" format may be called for. These things can generally be spotted during the development phase and can be addressed before the actual acceptance.
- Consider a provision that the company will review the website once a week as it is being developed. If the company does not object to anything that is evident in that review, the website developer can assume the company consents or agrees. This, or some variation, is potentially useful to both parties, but it requires the company to designate a knowledgeable person to do the weekly review. This "speak now or forever hold your peace" approach should not apply to the important specifications in the agreement where a more robust acceptance testing process should apply.

The acceptance testing process will commonly describe:

- How long the company has to test and review the website.
- The developer's obligations to correct problems with the website discovered through testing.
- The company's remedy in the event the developer fails to correct problems.
- When the website is considered accepted by the company (e.g., is deemed acceptance permitted or must the company accept in a written notice?).

Warranties

Warranties are an important element to any website development agreement. Common warranties found in these types of agreements include those described below.

- Intellectual property (e.g., website developer warrants that whatever intellectual property it contributed does not infringe, and company makes the same warranty for whatever intellectual property it provided).
- Website developer warrants that no "open-source" software will be used (or if this warranty cannot be made, the developer provides specifics and the company either consents or not).
- Company may ask for warranties relating to the experience or capability of the developer (or this may be covered in the recitals as noted above).
- Company may ask for warranties that all of the individuals that the website developer assigns to the project have invention assignment and confidentiality agreements with the developer (although the company may want to individually interview each such person and make sure that each person understands the relevant confidentiality issues).
- Company may ask that the developer warrant that it is appropriately qualified and licensed to do the work.
- Company may ask the developer to warrant that it is not aware of any aspect of the work that will infringe any other obligations—such as a warranty that the developer has not previously agreed to any noncompetition clauses that could be triggered by the work.
- Company may ask for a general warranty that the work will be done in a professional manner in accordance with standards in the industry and in a good workmanlike manner.

The warranty clause will almost universally contain a disclaimer of any implied warranties of merchantability or fitness for purpose, or any other warranty not expressly contained in the agreement.

Indemnifications

With respect to indemnification, claims relating to intellectual property and infringement are often those of most concern in a website development agreement. As with most other indemnification clauses in information technology agreements, it is important to also include the procedures with respect to indemnification. The indemnified party is typically required to:

- Give prompt notice of any claim;
- Allow the indemnifying party to defend the claim;
- Agree to cooperate in the defense of the claim (generally at the indemnifying party's expense); and
- Turn over copies of all correspondence or other documentation relating to the claim, and agree to continue to do so during the course of the dispute.

Sometimes it is appropriate to have a "basket" that limits the indemnification for matters that exceed a certain amount and, perhaps, a cap by which indemnification is limited to a certain maximum dollar amount. These types of terms are only appropriate under certain circumstances and should not be used as a general rule. The parties should also determine ahead of time whether payments to be made pursuant to, or damages that arise as a result of, a party's obligation to indemnify the other party are limited by the limitation of liability in the agreement, or whether such payments and damages are uncapped with respect to any exclusion of consequential damages or cap on direct damages set forth in the agreement.

Content of the Website

As part of its development process, the company will want to determine what content it wants on its website and in what form it is going to deliver that content to the developer. To some extent this may affect the developer's fees because the developer may find delivery in certain formats easier to use than other formats. Content examples may include:

- The company catalogue
- The company organization chart
- Other company promotional material

The website development agreement should also clearly articulate for what content of the website the developer will be responsible. Here are a few examples:

- Company wants a link to a database containing medical terms.
- Company wants a link to a database containing phone numbers.
- Company wants a link to a database allowing for stock quotes or other financial market information.

In addition to the foregoing, the website development agreement should state which party is responsible for securing the appropriate permissions and licenses for the content. If the website is to have professional advertising (i.e., ads with professional models, or where professional photos of the company products or people are included), the details of that should be specified in the agreement, and the appropriate responsibilities for obtaining the necessary releases and permissions established. Finally, ownership of all elements of the content should be clear. Virtually everything on the website should be subjected to the question, "Who really owns this?" If it was created by company employees, the ownership should be clear because the copyright laws provide for ownership by the company of anything created by its employees within the scope of their employment. However, this does not extend to material created by independent contractors nor, of course, to material that employees may have improperly taken from other sources.

Linking Issues

Linking to other content, websites, documents, and the like, is a common website practice that is used by many companies. When considering linking, it is essential that the parties in a website development transaction put in place the proper controls to ensure that the linking is done without violating another party's intellectual property or other rights.

- The parties should know, ahead of time, what consents are required for any links that the company wants to have on the website and who is responsible for obtaining the consents.
- If the developer is responsible for obtaining the appropriate consents, there should be a point at which these are turned over to the company for its permanent records.
- The parties should determine what notices or disclaimers should appear on the screen when one of the links is activated.

Insurance

Since much of the website development work might be done on the company premises, it is sometimes thought appropriate to have the website developer carry the same type of insurances that other nonemployees who work on the company premises are required to obtain—and possibly to furnish appropriate proofs of the existence of that insurance. Many standard website development agreements do not include insurance provisions, and because many website developers are solo practitioners or work for small companies, it may be difficult to ensure that such insurance coverage is obtained. Regardless, it is an important topic that should be part of a company's diligence process in selecting an appropriate developer.

Reports, Records, and Audits

It is common in website development agreements to have a reporting structure and to require that the developer maintain records associated with the services provided pursuant to the website development agreement.

- Consider what reports or records the website developer should create and provide to the company.
- Is the website developer authorized to incur expenses that the company will reimburse?
- What records are required for expense reimbursement and what should the audit rights be?

Training/Education/Troubleshooting

A company may require a website developer to train company employees on either the use of the website or how to make minor adjustments or minor content additions or deletions to the website. Details with respect to the training should be included in the website development agreement in some form—either as part of the main terms and conditions or as a separate training exhibit. The same is true for any instruction manuals or similar documents that the website developer may be expected to leave with the company for minor changes to the site. In some cases, it is appropriate to require that the developer provide a troubleshooting manual, which should be periodically reviewed by the company and subject to the same testing and acceptance procedures as the website itself.

Additional Provisions to Consider

- **Disputes.** Many website development agreements contain a form of alternative or informal dispute resolution.
- **Trademarks.** In many website development transactions, the developer will be provided with the company's manual or a set of requirements on the proper use of the company's trademarks and instructed that those specifications should always be followed. If that is the case and if there are requirements with respect to use of the company's intellectual property, consider also adding a process for how the developer's work be reviewed from time to time to ensure the proper use of the company's intellectual property.
- **Copyright.** Website development agreements frequently contain a process to ensure that the developer does not use anyone else's copyrighted material inappropriately. Further, the agreement typically describes the developer's obligations with respect to assigning to the company the copyrights to anything created for the company.
- **Privacy.** Most companies will have a specific privacy policy, and that policy (or at least a version of it) will go on the website in a certain place where it can be reviewed by visitors. Bear in mind that some states require websites to have a privacy policy. The website development agreement might include

a provision that the website developer will put a certain statement and type of link on the website for this privacy policy. The company should, of course, be responsible for creating the policy and deciding at least the general way in which the privacy statement will appear on the site.

- **Terms of use.** The company should assume responsibility for preparing whatever terms of use it feels are appropriate, and the website development agreement might simply call for the developer to put the appropriate statement and link on the site in the agreed-upon manner. The Terms of Use generally make clear the site is the property of the company, it may be used only for the visitor's own use, that the content is provided as-is, there are no warranties, and that the company's liability is strictly limited (usually to no liability for damages whatsoever).

Summary

While website development agreements will contain many of the same provisions found in other information technology professional services agreements, there are several critical areas that cannot be missed. These types of agreements can get very complicated as companies develop increasingly sophisticated websites and use cutting-edge and untested technology. While the provisions discussed in this chapter are essential, it is also important to include the common provisions found in other information technology agreements such as *force majeure* provisions, notice requirements, modification and assignment restrictions, and the like. It is also important to recognize that many website development agreements are really "combination" agreements involving the development of the website and other matters. These might include hosting, colocation, maintenance, and e-commerce services. In those types of combination transactions, it is important to avoid getting lost in the weeds and ensure that, while the other portion of the deal might be critical, the website development portion should not be lost and the provisions summarized in this chapter should be considered.

Chapter 26

Social Media Policies

CHECKLIST

Key Steps

- ☐ Understand your company's need for a social media policy
- ☐ Acknowledge that the definition of social media is a moving target—so define the term broadly
- ☐ Describe the overall scope of the social media policy and how it relates to other company policies
- ☐ Set guidelines for internal computer use generally; employees should know that:
  - If it is done on a work machine, it belongs to the company
  - Content produced on work machines is not private
  - Content may be monitored
- ☐ Set guidelines for outbound communications; employees should:
  - Exercise care in drafting all communications, whether personal or professional; pause before posting
  - Never post inappropriate content
  - In general, employees should not hold themselves out as representative of company when posting social media content.
  - Establish approval and moderating process for employees who wish to hold themselves out as representative of company (e.g., writing a professional blog)
- ☐ Employees should ask questions and ultimately sign the policy

Introduction

Social media (sometimes referred to as "social networking" or "Web 2.0") is an integral part of people's lives and has become a key marketing strategy for most businesses. People are participating in social media at staggering rates through the many social media outlets that now exist. Many companies have recognized the benefits that can be achieved through the use of social media in the workplace, including supporting the company brand, providing a forum for customer feedback, and building professional networks. The use of social media can help credential your business, connect with prospective clients and customers, and support professional development efforts. Furthermore, with the increasing number of daily interactions with social media, every company must be aware of how unfettered, uncontrolled use of social media by the company's employees—both inside and outside the work environment—can expose the company to undue risks. Those risks can be addressed and reduced through the adoption of an appropriate company social media policy. This chapter provides an overview of company social media policies and why such policies are important to a company and includes guidelines for development and implementation.

While what precisely constitutes "social media" is constantly debated by the media, technology industry professionals, and consumers, it is clear that social media uses web-based technologies to turn communication into interactive dialogues and into the creation and exchange of "user-generated content." The widespread use of social media in individuals' lives means there is an inevitable overlap with those individuals' places of employment, which subjects a company to legal risks. For example, an employee promoting the company's products may trigger Federal Trade Commission (FTC) scrutiny, making proprietary information public can have trade secret implications or violate a nondisclosure agreement with a third party, and discriminatory or violent statements may implicate harassment or other serious charges. Even if an employee does not intend to make his or her communications public, the ease with which information is shared and forwarded via social media technology may result in public dissemination of information that was intended to remain private. The risks may also be far more direct: a dissatisfied employee may post intentionally derogatory comments about the company, coworkers, or company leadership, damaging the company's reputation.

A social media policy outlines for employees the company guidelines and principles for communicating in the online world. The policy is not intended to prohibit employees' participation in social media on a personal level, but rather to provide guidelines regarding such participation in order to protect the company's interests and minimize potential risks. The policy should address the employees' use of company computer resources and provide guidelines regarding the employees' outbound communications. The policy should also be consistent with other company policies that limit employees' use of company computer resources in a manner that may adversely affect the availability of such resources to other users. Websites that use a large amount of bandwidth such as YouTube or Netflix may be blocked entirely or their use otherwise restricted. The policy should state that an employee participating in social media should not disclose confidential or otherwise sensitive or proprietary company information, nor should the employee imply that statements made by the employee are in any way representative of the company. Furthermore, the policy should bar any employee use of the Internet for non-work-related matters in a manner that negatively affects the employee's or any other employee's productivity. Finally, while the policy is primarily directed at company employees, the company should also consider including similar guidelines and restrictions in the policies it requires of its contractors and other third-party service suppliers. Examples of these provisions and other important provisions of a social media policy are discussed in detail below.

Some companies may also choose to allow employees to use social media as a means of promoting the company and its products and services. In that instance, the policy should require any use of social media as a formal company communication to fully comply with the company's applicable communications guidelines and require the employee to first obtain approval of the appropriate company representatives. The company should also consider having all participants in such formal communications register their user names with the company in order to facilitate monitoring of those communications.

Policy Scope and Disclaimers

While there is no "one size fits all" solution and each company must consider its particular situation, the following provisions are frequently part of comprehensive social media policies.

- The policy should clearly describe the scope of its reach. For example, the policy should state that, if applicable, it applies to multimedia, social networking websites, blogs, and wikis for both professional and personal use. Consider providing examples of the types of social media covered by the policy including Facebook, YouTube, Twitter, foursquare, MySpace, LinkedIn, and personal blogs, but don't limit the policy to just those social media websites listed.
- The company should reiterate that its policies related to trade secret and confidential information apply equally to information disseminated via a social media website.
- The policy should make clear that social media websites have their own terms of use, privacy policies, and other conditions. The policy should ensure that employees are made aware of their obligations to read, understand, and comply with those terms of use, privacy policies, and other conditions.

- The policy should state that it is not meant to interfere with the employees' rights as set forth in Section 7 of the National Labor Relations Act.
- If an employee comments on any aspect of the company's business, she must clearly identify herself as an employee and include a disclaimer such as "the views expressed are mine alone and do not necessarily reflect the views of (company's name)."
- All postings must respect copyright, privacy, fair use, financial disclosure, and other applicable laws.
- Company blogs, Facebook pages, Twitter accounts, etc. could require approval when the employee posts about the company and the industry.
- In most cases, the company should reserve the right to request that certain subjects are avoided, to withdraw certain posts, and to remove inappropriate comments.
- The policy should contain a statement that the technology resources and electronic communications systems belong to the company and are only to be used for authorized purposes.

  It should be noted that individuals designated as authorized by the company to speak on its behalf in social media forums require special treatment. Those individuals are "speaking on behalf of the company" and must be carefully trained regarding the policies of the company and statements made on its behalf. While social media frequently requires rapid responses online, the danger of granting an individual access to corporate social media accounts needs to be carefully addressed. Postings could violate laws, cause the company to be viewed in a bad light, and create consumer backlash. Wherever possible, a second set of eyes (a safety reviewer of content) should be used to ensure postings are appropriate before they are broadcast over the Internet.

No Expectation of Privacy

The privacy rights, if any, of the employees should be explicitly defined. It is generally the case that employees do not have (and should not expect) privacy rights with respect to any message or document created, stored, received, or sent with the company's technology or electronic communication resources. If applicable, the policy should state this clearly. In some instances, each time employees sign on to the company network, they are reminded that their e-mail is not private and that they do not have a reasonable expectation of privacy to the material created with computer resources. In those cases, it would be very difficult for an employee to maintain any reasonable expectation of privacy in his or her communications generated or received on the company's technology.

Right, But No Duty, To Monitor

If your company intends to monitor employee usage of social media sites, the policy should state that the employer has the right to monitor all manner of use of the company's technology and electronic communication resources. In these cases, the policy should explain that monitoring may include messages, including e-mail, texts, voicemails, posts, etc., that are stored, received or sent, in archival or other form. The policy should also provide that the employer is not obligated to conduct monitoring or review all messages or material, so that the policy does not create an affirmative duty on the employer to protect employees from inappropriate or abusive messages (whether e-mail, texts, posts, or anything else).

Conduct in Social Media

An employee's conduct in using social media and other electronic forms of communication can subject a company to significant risk. It is essential that social media policies clearly communicate the company's requirements with respect to employee conduct in these environments.

- Most policies in this area communicate to employees the unique nature of e-mail, texts, and other postings, specifically that such communications are easily copied and forwarded, are becoming a growing focus of litigation, and are difficult to delete with any permanence. Similarly, these types of policies impress upon employees that e-mail and text messages, as well as posts, reflect upon them as well as the company and that great care should be used in drafting all messages, whether e-mail, text, blog posting, postings to social media sites, or any other form.
- Many social media policies describe the types of inappropriate conduct that should be avoided. This can be very important as a result of employees using social media and other means of electronic communication for both business and personal use. Where lines of inappropriate conduct are drawn is often unclear.
- Consequences for creating, sending or forwarding messages, or postings with inappropriate content are usually detailed in social media policies. Such consequences should not be limited, but should allow for discipline up to and including termination.
- Each employee should be required to follow the law (e.g., all copyright, trademark, securities, privacy, and other laws) when posting content online. Employees are generally barred from posting content that could be characterized as defamation, plagiarism, a copyright violation, a securities law

violation, or that would otherwise violate applicable law. Employees should be encouraged to use citations or hyperlinks to original content when possible.

- Where applicable, the social media policy should remind employees that rules of professional conduct will continue to apply to their communications, including advertising and solicitation rules. Improper online pretexting should be avoided (e.g., attempting to gain access to information by posing as a confidant, or as one who is seeking a genuine social relationship with someone else to gain access and information).
- Real names should be used in social networking sites. The policy should discourage and, as appropriate, prohibit, the use of aliases or anonymity. The policy should also prohibit employees from misrepresenting who they are, what they do, or for whom they work. The policy should reflect that it is important to be honest and transparent in the online world.
- Social media should be avoided as a means to comment on or recommend individuals who are candidates for positions at the company. It is also not a good idea to communicate with candidates using social media during the recruiting process.

Social Networking and Weblogs

Social media policies generally describe the content that may be posted to social networking websites and blogs. The potential risks associated with such social media participation by employees should be emphasized and the policy should strictly delineate what activities and content are prohibited.

- Internet postings should not disclose any information that is confidential or proprietary to the company or to any third party that has disclosed information to the company.
- Postings should not include company logos or trademarks unless permission is asked for and granted. Employees should neither claim nor imply that they are speaking on the company's behalf.
- Postings should be limited to only those that employees would be comfortable having the company, colleagues, clients, and the general public read, hear, or see. Employees should be prohibited from posting anything that would potentially embarrass them or the company, or call into question their or the company's reputations.
- These types of policies generally make clear that postings are required to be accurate, truthful, respectful, and are checked for spelling, grammar, language, and tone. Employees should avoid language that would not be acceptable in the workplace. The same guidelines should apply to visual content—personal videos and pictures posted should be in good taste and support a positive personal brand.

- Social media policies should include rules with respect to setting up and maintaining personal blogs. For example, many companies prohibit employees from using company resources or technology in the creation or maintenance of personal blogs. Further, these policies generally prohibit an employee from using the name of the company in the name of the personal blog.

Employee Questions and Signature

Each employee allowed access to or use of the employer's technology or electronic communication resources should have an opportunity to ask questions about the policy, then should be required to execute an acknowledgment and agreement to comply with the company's technology and electronic communication use policy. This document should then be maintained in the employee's personnel file.

Summary

With the rapid growth and adoption of social media, a company must have an appropriate policy in place to provide guidelines to the company's employees in order to minimize the risks associated with those employees' use of social media. The provisions described in this chapter can be used to create a robust social media policy that will reduce the risk to the company associated with employee's use of social media and will educate the employees on the dangers of using social media for work and other purposes. Each of the company's policies, including the social media policy, should be consistent with the company's other policies related to communications, confidentiality, and use of company resources. In addition, these types of policies work best when they reflect the company's culture and values. As social media continues to grow and evolve, the social media policy may need to be changed to reflect the new landscape.

Chapter 27

Critical Considerations for Records Management and Retention

CHECKLIST

Scope

- ☐ Applicability to all types and formats of records
- ☐ Applicability to all affiliates, divisions, and business units
- ☐ Risk assessment
- ☐ Third-party contractors and outsourcers
- ☐ Active vs. inactive records

Retention Schedule

- ☐ Detailed list of records categories
- ☐ Employee surveys and interviews
- ☐ Organized by department/business unit
- ☐ Retention periods based on applicable law
- ☐ Retention periods based on operational needs
- ☐ Retention periods based on statutes of limitation
- ☐ Citations to applicable laws
- ☐ Periodic (e.g., annual) review and update

Litigation Holds

- ☐ Responsibility for issuing litigation holds
- ☐ Litigation hold notice
- ☐ IT department involvement in litigation hold process
- ☐ Notification of outside vendors and outside counsel
- ☐ Employees obligated to notify management of pending or foreseeable claims
- ☐ Termination of litigation hold
- ☐ E-discovery procedures
- ☐ Data map

Electronic Records

- ☐ Authorized storage locations
- ☐ Retention, archiving, and destruction of e-mails
- ☐ Retention, archiving, and destruction of voicemail
- ☐ Security and encryption where required (e.g., protected health information, sensitive financial information, laptops and removable media)

Administration

- ☐ Designated records manager
- ☐ Input and approval by board and senior management
- ☐ Confidentiality of employee personnel and medical records
- ☐ Approved methods for destroying paper and electronic records
- ☐ Procedures for distribution to and training of employees
- ☐ Auditing compliance with the policy
- ☐ Off-site storage of inactive paper records

Overview

In light of the vast volume of computer and other electronic files and communications, and litigation obligations with respect to e-discovery, companies now realize the need for a comprehensive records retention policy. Failure to retain records in compliance with applicable law and in connection with pending or threatened claims can result in regulatory and court sanctions, fines, unnecessary expense, and other adverse consequences. Inadequate and ineffective records storage and retention practices can result in (i) the loss of valuable trade secrets, confidential information, and other important business and proprietary information, and (ii) the breach of privacy laws and regulations. The cost (time, money, and resources) of complying with litigation discovery requests can be significantly reduced through implementation of cost-effective records retention and e-discovery policies and practices.

Keeping everything is not the answer. Companies that say they keep everything typically do not. Employees will always discard paper records, electronic files, and e-mails. Destroying records without a policy can result in inconsistent and haphazard retention and destruction practices frowned upon by courts. In light of this inevitable destruction of records, it is imperative for the company to adopt a policy governing destruction in order to avoid liability for selective destruction of records or spoliation.

The benefits of an effective records management program include easier and timely access to necessary records; complying with statutory and regulatory retention obligations; reducing storage costs; protection of confidential and proprietary information; and meeting e-discovery obligations. An effective records retention policy can mitigate the risks of not actively managing electronically stored information (ESI), such as the inability to efficiently locate and use important business information, sanctions due to the failure to comply with statutory and regulatory retention and destruction laws, increased costs due to inefficiencies from inaccessible information, and the inability to comply with e-discovery requirements, court orders, and other litigation-related requirements.

Avoiding Spoliation Claims

"Spoliation" is the term commonly used by courts to describe the improper destruction of evidence, most typically documents and records. Although the precise rule varies slightly in different jurisdictions, generally a party is guilty of spoliation if it destroys evidence (e.g., company records) relevant to litigation with the purpose or intent of preventing the other party from using the evidence against the party. Liability can arise from destruction prior to litigation being instituted and sometimes even in the absence of an actual threat of litigation.

The consequences to the "spoliator" can be dramatic. Remedies for improper destruction of records can include (i) monetary sanctions or penalties, (ii) an inference in the litigation permitting the jury to presume that the documents contained damaging information (i.e., information supporting the other party's position), (iii) sanctions ordering that certain facts be deemed established or preventing the spoliator from opposing a certain factual assertion, (iv) dismissal of claims or entry of default judgments, and (v) criminal liability. In light of the severe consequences of improper destruction, businesses must take a proactive approach to drafting, implementing, and enforcing their record retention policies.

Impact on Litigation/Discovery Costs

Companies that have invested the time and resources to prepare a comprehensive records retention policy can comply with their discovery obligations efficiently. In contrast, companies that do not prepare in advance have found themselves unable to make required disclosures and to timely comply with discovery obligations without incurring tremendous costs. Most significantly, companies who have not prepared for e-discovery have suffered evidentiary and monetary sanctions.

As noted by Judge Shira Scheindlin of the Southern District of New York in the first of the seminal Zubulake decisions, "the more information there is to discover, the more expensive it is to discover all the relevant information until, in the end, 'discovery is not just about uncovering the truth, but also about how much of the truth the parties can afford to disinter'" *Zubulake v. UBS Warburg*, 217 F.R.D. 309 (S.D.N.Y. 2003); quoting *Rose Entm't, Inc. v. William Morris Agency, Inc.*, 205 F.R.D. 421, 423 (S.D.N.Y. 2002). An effective records retention policy can assist companies in dealing with the tremendous volume of electronic records and reducing the costs of complying with electronic discovery requests.

Developing the Policy

In developing a records retention policy, the company should first analyze the records environment to assess areas and levels of risk to the organization that may result from existing records retention policies and practices. Based on identified risk areas, the company can then evaluate existing written and/or de facto policies, processes, and technologies to identify weaknesses, categorize risks, and recommend improvements. With the results of the risk and needs assessment in hand, the organization can then modify the existing policy or develop a new, practical, and cost-effective records management and retention policy that addresses and resolves any potential issues revealed during the risk and needs assessments.

Litigation Discovery Procedures

To avoid court sanctions, costly e-discovery compliance, and missing court deadlines, companies should prepare in advance to properly respond to e-discovery requests and mandatory disclosures. The company should provide training to its personnel with respect to the policy to assist compliance with paper and e-discovery obligations. A critical aspect of litigation preparedness is knowing what electronic records the company maintains and where they are stored. The company should develop legally compliant data maps that categorize the company's electronic records and identify where the records are stored, as well as the appropriate records custodians who can provide electronic records as needed.

The records retention policy should address obligations under the federal and state rules of civil procedure relating to electronically stored information. These rules require early treatment of e-discovery issues, as well as full and accurate disclosure of the existence of relevant ESI. If not properly planned, managed, and coordinated, locating and producing ESI can become very time-consuming and expensive. Failure to comply with the discovery rules can result in court-imposed sanctions, fines, and adverse rulings. Accordingly, it is critical for companies to develop accurate documentation describing their ESI practices and policies on the "front end," rather than dealing with these issues on an ad hoc, case-by-case basis after litigation has commenced.

Developing the Retention Schedule

An essential component of every records retention policy is the retention schedule that identifies all different types and categories of records and the required retention periods. The retention periods may be based on a statute, regulation, or other law that mandates the record be retained for at least a specified period of time. In the absence of a legally mandated retention period, operational requirements will dictate how long records should remain available. Failure to utilize an accurate retention schedule can lead to premature destruction of records, resulting in legal fines and sanctions and loss of information needed for the ongoing operations of the business. The company's records retention policy should have a retention schedule that accurately and concisely identifies all different categories and types of paper and electronic records retained by the company and legally compliant retention periods for each category or type of record.

- **Compliance with applicable law.** First and foremost, retention periods must comply with applicable laws. Many types of records have mandatory retention periods required by law. For example, in the area of human resources, federal law requires payroll records to be retained for at least three years, and records relating to employment actions (e.g., hiring, promotions, demotions, firing) must be retained for at least one year. (26 C.F.R. 1627.3; 29 C.F.R. 1602.14; 29 C.F.R. 516.5.) In the area of employee safety, employee medical records and records relating to exposure to hazardous materials must be retained for at least thirty years. (29 C.F.R. 1910.1020.) State laws and regulations also contain record retention requirements.
- **Senior management involvement.** Senior management must stress the importance of developing the records retention policy and an accurate schedule in order to maximize the likelihood of receiving timely and accurate responses to the surveys. If possible, it is preferable to provide applicable employees with a sample or starting list. The starting list might come from existing lists within the organization that need updating or from legal counsel or consultants with records retention expertise.
- **Litigation hold.** Records that are relevant or may be relevant to pending, threatened, or reasonably foreseeable litigation, claims, investigations, or

other legal proceedings must be preserved from destruction. If the organization becomes aware that a lawsuit or other legal process has commenced or is likely to commence, records relating to that claim may not be destroyed. The records should be identified, segregated, and preserved until the claim is resolved. This is often referred to as a "legal hold" or "litigation hold." The legal hold process should be coordinated by the company's legal counsel—whether in-house, external, or both. Companies that do not avoid destruction of records subject to a legal hold scenario face risks of sanctions, including monetary fines, adverse rulings in court, and other detrimental consequences.

- **Data collection**. The first step in developing the retention schedule is data collection. If the company has an existing retention schedule or list of records, they can be used as a starting point in the development of the retention schedule. Often, a company will not have an enterprise-wide retention policy in schedule; however, certain business units or departments may have developed their own schedules out of necessity or as a useful tool.
- **List of records categories**. The records list should be detailed enough to accurately reflect and capture all different types and categories of records maintained by the company, however, should not be so long and detailed as to render the list meaningless or overly difficult to use. Remember, the goal is to have a defensible policy that can and will be substantially complied with by the employees of the organization.
- **Surveys**. The records list is typically prepared by using written surveys, interviews of key personnel, or a combination of both. The survey is essentially a questionnaire designed to solicit information and identification of the different types of records maintained by a particular department or business unit. It typically is a waste of resources to survey all employees. Accordingly, the company must decide which employees should receive the survey. Department managers and leaders typically are in a good position to identify the appropriate personnel to receive the surveys.
- **Interviews**. Interviews are also useful in collecting the necessary information to develop the records list. Interviews have the advantage of one-on-one communication, which allows for interaction, questions, and answers in a format more efficient than the survey process. Additionally, the interview has the added benefit of "getting the job done" rather than the inevitable delays and back-and-forth that may result from using a survey. Finally, interviews can be more easily tailored to the specific personnel and business function. As with the survey process, the company must carefully determine the most appropriate individuals to be interviewed.
- **Structure or retention schedule**. For most organizations, the best way to structure the retention schedule is by using the internal organization structure used by the company, for example, by departments, divisions, business units, and/or business lines. Thus, employees in the human resources department will know to first look at the "Human Resources" section of the schedule,

while accounting personnel will be directed to the "Accounting" or "Finance" section of the schedule. The schedule should be published in an electronic format (e.g., Word or .pdf) so that it is word searchable, which will make it easier for employees to locate the applicable retention periods.

- Keep in mind that a department-based schedule can result in duplication. For example, many different departments may deal with invoices, contracts, and inventories. While there may be legitimate business reasons to retain similar records for different periods of time depending on the business function, the same or similar records should be retained for consistent time periods as much as possible.

- **Determining retention periods**. After developing the records list, it is necessary to determine how long each type of record should be retained. The first question to ask is whether a federal or state law, regulation, or statute requires retention for a particular period of time. If so, this will be the minimum retention period. Most records, however, will not have a clear and explicit legally read wired retention period. Accordingly, the organization will need to consider other factors in determining the retention period, for example, ongoing business operations, litigation, regulatory audits, customer service, budgeting, and strategic planning.
  - The number of different retention periods should vary (e.g., one year, four years, seven years), but having too many different retention periods adds to the complexity of the schedule and makes compliance more difficult. Accordingly, many companies will use three or four "buckets" such as short-term (one year), medium-term (three years), long-term (seven years), and very long-term (e.g., thirty years for OSHA- and hazardous waste–related records).
  - In determining the retention period, the company should ask not how long a record is kept, but rather, being realistic, how long the record is really needed. Many employees overestimate the value of their records and mistakenly believe they should be kept longer than they really should be retained. Similarly, employees often overestimate the risk of destroying records. Thus, personnel involved in determining retention periods should be educated on the importance of retaining records long enough, but not too long.
  - One common factor in assessing appropriate retention periods is whether the records are likely to be relevant in the event of litigation. For example, contracts, accident records, customer complaints, research and development, and employee disciplinary records are often relevant and needed for litigation, whether for prosecuting or defending claims. For these types of records, consideration should be given to the applicable statutes of limitation. For example, if the statute of limitation for contract actions is five years, contracts and records relating to the contract should be retained for at least a period of five years after termination of the agreement.

- Keep in mind, however, that a statute of limitation relates to how long a party can wait before bringing a lawsuit and is not an explicit retention period. Thus, a company with operations in several states may choose a reasonably long retention time based on the various statutes of limitation; however, not necessarily the longest statute of limitation. For example, Illinois has a ten-year statute of limitation for written contracts. Six years or less is a common limitation period for written contracts. A company with operations in Illinois and other states with shorter limitation periods may appropriately use a six-year retention time for records retention based on statutes of limitation. Most lawsuits are brought within a year or two of the circumstances giving rise to the claim. As time passes after a particular event, the likelihood of a lawsuit diminishes. Thus, a retention period shorter than a long statute of limitation may be appropriate under the circumstances.
- Retention periods generally come in two types—the time may be based on when a document is created, or when the document is no longer "active." A contract is active for the period of time it is in effect. For example, a company policy is active until it is superseded by a new policy. A contract is active until it is terminated or expires. Thus, a retention period for an invoice may be "1"—meaning it should be retained for one year after it is created. A retention period for a contract may be "A+5" meaning it should be retained for five years after termination of the agreement.
- Generally, it is not advisable to use "permanent" or "indefinite" retention periods. When used, permanent typically means the record is kept "forever." Indefinite records do not have a definitive retention period, but should be periodically reviewed to determine if they are eligible for destruction and are not subject to continued legal or business retention requirements. Overuse of the permanent retention period is discouraged as it defeats one of the primary benefits of a retention policy, namely the ability to destroy records. Indefinite times are problematic as they add subjectivity and uncertainty to the records retention and destruction decisions. Subjectivity can result in inconsistent and haphazard retention practices, increasing the risk of loss and liability due to untimely or improper destruction of records.
- A company's past experience should also be considered in determining retention times. For example, a company that experiences a high level of litigation relating to a particular business function should examine past history in determining which types of records have been relevant or needed for the litigation, and determining the retention time, most likely driven by the applicable statute of limitation, and secondarily, the typical timing of when claims are filed.
- Numerous resources exist for identifying laws, statutes, and regulations that contain mandatory retention times. Nonetheless, the number of

such laws and regulations is so numerous, particularly when including the state level, it is not always a simple or quick task. Legal counsel and consultants with experience and expertise in records retention matters will likely have existing work product and databases with legally required retention times. Commercially available legal databases, such as Westlaw and LexisNexis, can be utilized for electronic searching. Additionally, Information Requirements Clearinghouse (www.irch.com) provides useful records retention resources.

- Citations to the applicable laws and regulations should be identified either in the retention schedule itself or in separate documentation maintained as part of the records retention program. This is particularly useful if it is necessary to review the actual law to determine applicability to a particular type of record. Additionally, tracking the applicable laws makes periodic updating to comply with changes in the law and confirmation of the retention times easier and more efficient.

The E-Mail Problem

E-mail proliferation is a problem faced by every company. Confronted with growing storage costs and system performance issues, companies are limiting the amount of e-mail that employees can keep. Tape or other backups are "snap shots" of data at a particular time and do not keep a complete record of all e-mails. While limiting e-mail volume is legally appropriate, and in many cases advisable, the company must also ensure employees (or an automated system) do not delete e-mails that are required for ongoing business operations or legal compliance. Companies should implement policies and practices for ensuring required e-mails are not prematurely destroyed, for example, by migrating or archiving required e-mail records to a document management system or secure networked data servers.

Authorized Storage Locations

Electronic records can be stored in a variety of locations—network servers, local hard drives, home computers, laptops, handheld devices, smart phones, CD-ROMs, flash storage devices, web-based e-mail applications, and online backup sites. Multiple locations add to the difficulty and cost of locating and producing records and increase the likelihood that records will be lost, not produced when they should be, and/or improperly disclosed to third parties not entitled to access the records. When a company is required to locate and produce electronic records in litigation (as a party or a third-party witness), it must search all locations for potentially relevant records and produce those records. Companies should require storage or records in locations and in manners that facilitate prompt and cost-effective location and production and consider limiting the locations where electronic records may be stored by employees.

As noted above, electronic records should be stored only in company-approved and controlled locations. The next step is to create a data map or inventory of where all electronic records and other ESI are stored (e.g., file servers, e-mail servers, identified drives, storage networks, and removable media). This is important to facilitate the company in locating electronic records when they are needed for litigation or other legal proceedings. The data map is also critical to complying with the e-discovery rules, which require early and proactive disclosure of electronic records and information regarding their location and accessibility.

Confidentiality and Security

Electronic records often contain sensitive information valuable to the company, such as trade secrets, financial data, business plans, and other confidential business information. Similarly, with the increase of legislation regulating privacy of personal information, companies are under increasing obligations to maintain the privacy of such data. Accordingly, the company should implement and enforce policies and practices that protect the confidentiality, integrity, and security of important business information and adequately protect the privacy of personal information.

Third-Party Vendors

Many companies use independent contractors and outsource functions and operations of the business, resulting in third parties having primary responsibility for storing, retaining, and disposing of company records. Outsourced functions include areas such as information technology, accounting, human resources, or other business processes. In such instances, the company should require the outsourcer to comply with the company's records management policies through appropriate contract language, monitoring, reporting by the outsourcer, and periodic auditing of the outsourcer.

Proper Destruction

The flip side of retention is destruction. In order to obtain the benefits of having a policy and avoiding liability for improper destruction of records, it is necessary to destroy records in accordance with the policy. The company should regularly destroy records in accordance with its policy, subject to suspension of destruction pursuant to a litigation hold. Records destruction is typically undertaken on a periodic basis, such as annually. A general identification of what records are destroyed should be maintained. Additionally, records containing confidential or sensitive information, such as health information or financial information, should be destroyed in a manner maintaining the confidentiality of the records. For example, such records should be shredded rather than simply thrown out with the trash.

It is not uncommon to require vendors and other third parties to destroy and erase company records according to a particular standard or industry best practice. Common language is along the lines of the following: In the event any hardware or storage media must be disposed of or sent off-site for servicing, provider shall ensure all client confidential information has been "scrubbed" and irretrievably deleted from such hardware and/or media using methods consistent with best industry practices (i.e., at least as protective as the DoD 5220-22-M Standard, NIST Special Publication 800-88, Guidelines for Media Sanitization, or NAID standards).

Summary

By instituting reasonable and appropriate measures, such as those described in this chapter, businesses can achieve better compliance with applicable document retention laws and regulations and can better protect their valuable proprietary information. These measures can also greatly reduce the costs of discovery in litigation.

Glossary

Actual Uptime: A measurement period, with respect to any particular service level measured in terms of availability, the aggregate amount of time within the scheduled uptime for such service level that the particular service being measured by such service level is available for use.

Additional Resource Fees or ARC: The incremental charges for each resource unit consumed above the applicable monthly resource baseline.

Affiliate: Any corporation, partnership, limited liability company, or other domestic or foreign entity (i) of which a controlling interest of more than 50% is owned directly or indirectly by a Party, or (ii) controlled by, or under common control with, a Party.

Aggregated Data: Refers to data that has been scrubbed of any personally or entity identifiable information and then generally combined with similar information from other parties.

AICPA: The American Institute of Certified Public Accountants.

Application Programming Interface or API: A specification for establishing interfaces between two or more elements of software, hardware, or other systems.

Approve or Approval: Those categories of approval specifically described therein, (i) the written authorization by a client's Chief Information Officer (or his or her designee) or an Outsourcing Relationship Executive (or his or her designee) for any consent, authorization, amendment, or other approval required under an agreement, and (ii) with respect to any consent, authorization, amendment, or approval requiring the authorization of payment, or imposing an obligation on the client for any fees, costs, or other expenses (a) the written authorization of client's Chief Information Officer; (b) the written authorization of client's Outsourcing Relationship Executive as provided in the agreement; or (c) the written authorization of the applicable client Functional Service Area Manager as provided in the agreement.

ASCII: An acronym for American Standard Code for Information Interchange. An almost universally accepted format for exchanging text-based information. ASCII format is, however, limited in that it does not preserve the formatting of the text or any special characteristics of the document (for example, footnotes, tables, and bullet points).

ASP: Acronym for Active Series Pages.

Audit Trail: An automatic feature of computer operating systems or certain programs that creates a record of transactions relating to a file, piece of data, or particular user.

Auditor: A Party's internal or independent third-party auditors, such independent third-party auditors designated by a party in writing from time to time, in its sole discretion.

Authentication: Verifying the identity of a user, process, or device, often as a prerequisite to allowing access to resources in an information system.

Authorized Users: Any individual or entity authorized by client to use the services under the agreement, whether on-site or accessing remotely.

Availability: The Actual Uptime expressed as a percentage of the Scheduled Uptime less Excused Downtime for such Service [i.e., Availability% = ((Actual Uptime/(Scheduled Uptime − Excused Downtime)) × 100)].

Available for Use: The ability of equipment, software, applications, and data, which are used to provide the services and for which provider is operationally responsible, to be utilized or accessed by authorized users in accordance with normal operations and without material degradation of performance.

Backups: Duplicate copies of data, generally stored at an off-site, secure facility.

Bankruptcy Code: Title 11 of the United States Code.

Batch: The entry of any defined batch header information, all of the records contained in the batch, footing of the totals, corrections found while footing to the totals, final balancing of the batch detail to the totals, and release of the information for use in the associated production system.

Billable Project: A discrete unit of nonoccurring work that is not (i) an inherent, necessary, or customary part of the day-to-day services in any functional service area, and (ii) required to be performed by provider to meet the existing service obligations or service levels (other than service levels related to billable project performance).

Bit: The smallest unit of data. A bit can have only one of two values: "1" or "O." See also **Byte.**

Byte: A basic unit of data. A byte consists of eight bits and can represent a single character such as a letter or number. A "megabyte" refers to a million bytes of information. A "gigabyte" refers to a billion bytes of information.

Cache: Memory used to store frequently used data. With regard to the Internet, caching refers to the process of storing popular or frequently visited websites on a hard disk or in RAM so that the next time the site is accessed, it is retrieved from memory rather than from the Internet. Caching is used to reduce traffic on the Internet and to vastly decrease the time it takes to access a website.

Central Processing Unit: Abbreviated "CPU." The portion of a computer that controls the processing and storage of data.

Certificate: A digital representation of information, which at least: (i) identifies the certification authority issuing it, (ii) names or identifies its subscriber, (iii) contains the subscriber's public key, (iv) identifies its operational period, and (v) is digitally signed by the certification authority.

Change: Typically any change (whether to software, equipment, services, interfaces, the system or any other related network, service, system, or hardware) that would alter the functionality, performance, or technical environment of the software, interfaces, the system or the equipment, the scope or manner in which the services are provided, the composition of the services, or the cost to the customer of the services or deliverables.

Click-Wrap Agreement: An agreement that is presented to the user for acceptance by clicking on an "I Accept" or similar means. The agreement is usually presented to the user as part of the installation process for a piece of software or as part of the registration process when a user is accessing an online service.

Client Computer: A personal computer or workstation connected to a network file server. See also **File Server**.

Client Data: All of the Client's data, records, and information to which provider has access, or is otherwise provided to Provider, that is entered into, transmitted by, or transmitted through the client's IT environment (including any modifications to any such data, records and information, any derivative works created therefrom, and any sorting routines applied thereto) under the agreement in connection with providing the services. Client's data typically excludes provider's confidential information or other provider intellectual property.

Client Equipment: The servers, hardware, machines, and other equipment owned, leased, or otherwise obtained by client as of the effective date and utilized by provider to provide the services.

Client-Server Network: A type of network in which server computers provide files to client computers. See **Client Computer** and **File Server**.

Client Third-Party Vendor: Any third-party vendor (other than provider or any provider third-party vendor) contracting directly or indirectly with client to provide any products or services.

Client Work: Tangible and intangible information and developments that are owned by client, including all intermediate and partial versions thereof and all designs, specifications, materials, program materials, software, flowcharts, outlines, lists, compilations, manuscripts, writings, pictorial materials, schematics, other creations, and the like.

Cloud Computing: A delivery model for IT resources and services that uses the Internet to provide immediately scalable and rapidly provisioned resources as services using a subscription- or utility-based fee structure.

Common Element: The commonly used elements that are not unique to client deliverables or unique client specifications including icons, program flows, displays, program elements, subroutines, algorithms, formulae, data structures, and other specifications.

Compliance: Conformity in fulfilling official requirements.

Compressed Files: A file whose contents have been "compressed" using specialized software so that it occupies less storage space than in its uncompressed state. Files are typically compressed to save disk storage space or to decrease the amount of time required to send them over a communications network like the Internet.

Consequential Damages: Are damages that are not a direct result of an act, but a consequence of that act. Consequential damages must be foreseeable at the time the contract is entered into. In connection with a breach of contract, consequential damages would include any loss the breaching party had reason to know of and which could not reasonably be prevented by the nonbreaching party. Consequential damages can include loss of business, loss of profits, and harm to business reputation.

Contract Band: A symmetrical range of usage above and below the Resource Baseline to which, with the exception of the Dead Band range, ARCs and RRCs apply.

Contract Year: Any one (1) year period ending at the end of the day before an anniversary of the effective date or, in the case of the last contract year, that portion of a year beginning on the expiration of the previous contract year and continuing until the completion of the term.

Contractual Privity: See "privity".

Cookie: A cookie is a small data file that a website can store on a visitor's computer. If the visitor returns to the website, the cookie can be used to identify the visitor and to provide personalized information to the visitor. Cookies are used by the operators of websites as marketing tools to gain information about their visitors and to track their movements on the site. Web browsers can be configured to reject cookies when they are offered.

Copyright Act: The United States Copyright Act of 1976 (17 U.S.C. §§ 101, *et seq.*), as amended.

CPI: Acronym for Consumer Price Index.

CPU: Acronym for Central Processing Unit. See **Central Processing Unit**.

Cross-over Issues: Issues related to the services that may cross over from one functional service area or provider to another.

Data Agreement: A license agreement for a database of information typically gleaned from public sources.

Data Feed Agreement: A type of data agreement where the data is constantly changing and made available through a continuous electronic feed (e.g., data relating to the financial markets).

Deal Memo: A memorandum used internally within a business to describe a proposed transaction.

Deliverable: An item or a service to be provided by a supplier under an agreement or statement of work that is identified as a deliverable, by designation or context, in a statement of work, exhibit, schedule, or other contractual document.

Direct Damages: Direct damages are intended to place the nonbreaching party in the position it would have occupied had the breaching party performed as promised under its contract. They are generally the difference between the value of the performance received and the value of the performance promised as measured by contract or market value. They are not intended to punish the breaching party.

Disabling Device: Any virus, timer, clock, counter, time lock, time bomb, adware, spyware, or other limiting design, instruction, or routine that would erase data, render data inaccessible, render data accessible in a modified format, erase programming, or cause any resource to become inoperable or otherwise incapable of being used in the full manner for which such resource was intended to be used.

Disk Mirroring: A method of protecting data from a catastrophic hard disk failure. As each file is stored on the hard disk, an identical, "mirror" copy is made on a second hard disk or on a different partition of the same disk. If the first disk fails, the data can be recovered instantly from the mirror disk. Mirroring is a standard feature in most network operating systems.

Documentation: All policies and procedures relating to the services and all training course materials (including knowledge transfer and computer-based training programs or modules), technical manuals, logical and physical designs, application overviews, functional diagrams, data models, production job run documents, specifications, reports, or other written materials identified, provided, or developed under the Agreement (as to each, whether in hard or soft copy).

ECI: Acronym for Employment Cost Index.

EDI: Acronym for Electronic Data Interchange. See **Electronic Data Interchange**.

Electronic Data Interchange: Abbreviated EDI. A method of electronically communicating standard business documents (for example, purchase orders, invoices, and shipping receipts) between the computer systems of businesses, government organizations, and banks. EDI was developed in the 1960s as a means of accelerating the movement of standard documents related to the shipment and transportation of goods.

Emergency Security Service Request: A request made directly to the provider's service desk by phone that requires immediate assignment to a security engineer for action. These requests may include firewall rule changes, account modifications (e.g., revoking access), remote access issues, digital certificate/encrypt ion issues, and team member suspicious activity log data requests.

Encryption: A method of using mathematical algorithms to encode a message or data file so that it cannot be understood without a password.

Environmental Laws: All applicable federal, state, and local statutes, laws, regulations, rules, ordinances, codes, licenses, orders, and permits of any governmental entity relating to environmental matters, including the following: (i) the Clean Air Act (42 U.S.C. §§ 7401 *et seq.*), the Comprehensive Environmental Response, Compensation and Liability Act of 1980 (codified in various sections of 26 U.S.C., 33 U.S.C., 42 U.S.C. and at 42 U.S.C. §§ 9601 *et seq.*), the Federal Water Pollution Control Act (33 U.S.C. §§ 1251 *et seq.*), the Federal Insecticide, Fungicide and Rodenticide Act (7 U.S.C. §§ 136 *et seq.*), the Safe Drinking Water Act (42 U.S.C. §§ 300f *et seq.*), the Toxic Substances Control Act (15 U.S.C. §§ 2601 *et seq.*), the Endangered Species Act (16 U.S.C. §§ 1531 *et seq.*), the Emergency Planning and Community Right-to-Know Act of 1986 (42 U.S.C. §§ 11001l *et seq.*), the Resource Conservation and Recovery Act (42 U.S.C. §§ 6901l *et seq.*), the Superfund Amendments and Reauthorization Act of 1986 (codified in various sections of 10 U.S.C., 29 U.S.C., 33 U.S.C., and 42 U.S.C.) and all regulations promulgated under any of the foregoing federal laws; and (ii) all other federal, state, and local laws, regulations, and ordinances similar in substance or intent to the laws described in the foregoing clause (i).

Excused Downtime: The aggregate amount of time in the month during Scheduled Uptime during which the applicable Service is not Available for Use by Authorized Users due to scheduled outages Approved in advance. For the avoidance of doubt, Excused Downtime shall not include time spent by Provider seeking the assistance of Third-Party Vendors or internal resources for Service Requests, Incidents, or Problems associated with Service Requests, Incident, or Problem Resolution.

Extranet: An extension of the corporate intranet over the Internet so that vendors, business partners, customers, and others can have access to the intranet. See also **Intranet** and **Internet**.

Feedback: Ideas, suggestions, etc., that a licensee may provide to a licensor regarding the licensor's products and services. Sometimes used interchangeably with **Submissions**.

Feedback Clause: A provision in a technology contract that makes clear that any suggestions or recommendations (i.e., "feedback") made by the customer about the vendor's products and services may be used by the vendor without limitation.

FFIEC: An acronym for the Federal Financial Institutions Examination Council, a formal interagency body empowered to prescribe uniform principles, standards, and report forms for the federal examination of financial institutions by the Board of Governors of the Federal Reserve System, the Federal Deposit Insurance Corporation, the National Credit Union Administration, the Office of the Comptroller of the Currency, and the

Office of Thrift Supervision, and to make recommendations to promote uniformity in the supervision of financial institutions. In 2006, the State Liaison Committee (SLC) was added to the Council as a voting member. The SLC includes representatives from the Conference of State Bank Supervisors, the American Council of State Savings Supervisors, and the National Association of State Credit Union Supervisors.

Field(s): Individual entries or groups of entries within a file relating to the same subject. For example, a litigation support database may have fields for the creator and recipient of a document and its subject.

File: A collection of data or information stored under a specified name on a disk. Examples of files are programs, data files, spreadsheets, databases, and word-processing documents.

File Server: A central computer used to store files (for example, data, word-processing documents, programs) for use by client computers connected to a network. Most file servers run special operating systems known as "network operating systems (NOS)." Novell Netware and Windows NT are common NOS. See also **Client Computer** and **Client-Server Network**.

Fixed Fee: A professional services engagement in which the contractor will perform services for a specified fee, regardless of the actual time and expense required to complete the work.

Force Majeure: Force majeure clauses are common in technology contracts. The term literally means "greater force." These clauses excuse a party from liability if some unforeseeable event beyond the reasonable control of that party prevents it from performing its obligations under the contract.

GIF: Acronym for Graphics Interchange Format. A graphics file format used to exchange images on the Internet.

Hard Disk: A storage device based on a fixed, permanently mounted disk drive. Hard disks can be either internal or external to the computer.

HIPAA: The Health Insurance Portability and Accountability Act of 1996 and its implementing regulations.

Home Page: Generally the first of a collection of HTML pages, collectively forming a website. See also **HTML** and **Website**.

HTML: Acronym for Hypertext Markup Language. The formatting and layout language used to create documents for viewing on the World Wide Web. HTML tells web browsers how documents on the web are to be displayed.

Idea Submission Agreement or Submission Agreement: An agreement under which a third party, usually a customer or business partner, submits ideas, suggestions, and feedback to a company about its products and services.

IM: Acronym for Instant Messaging.

Incident: Any event or occurrence that is not part of the standard operation of a service, and that causes, or may cause, an interruption to, or a reduction in, the quality of that service. For the avoidance of doubt, excused downtime of an application or other IT infrastructure shall not constitute an incident. Also a service request shall not constitute an Incident.

Incident Resolution Time: The elapsed time between the time the applicable Incident is opened and logged into provider's system and the time a provider technician logs the incident as "service restored" (by repair or workaround) in that system, provided that an identical incident (i.e., same root cause) opened within one (1) business day after an incident is reported as "service restored" will restart the elapsed time. Incident resolution time will exclude, (i) time during which the applicable device is not remotely accessible, and (ii) if physical access to the applicable device is necessary to resolve the incident, time during which the applicable device is not available to provider's technicians at a customer service location.

Infrastructure as a Service (IaaS): The capability provided to the customer as to provision processing, storage, networks, and other fundamental computing resources where the consumer is able to deploy and run arbitrary software, which can include operating systems and applications. The consumer does not manage or control the underlying cloud infrastructure but has control over operating systems, storage, deployed applications, and possibly limited control of select networking components (e.g., host firewalls).

Instant Messaging: A technology (usually over the Internet) that permits individuals to chat online with others using a text-based interface.

Intellectual Property (IP): All inventions (whether or not subject to protection under patent laws), works of authorship, information fixed in any tangible medium of expression (whether or not subject to protection under copyright laws), moral rights, trademarks, trade names, trade dress, trade secrets, publicity rights, know-how, ideas (whether or not subject to protection under trade secret laws), and all other subject matter subject to protection under patent, copyright, moral right, trademark, trade secret or other laws, including, all new or useful art, configurations, Documentation, methodologies, best practices, operations, routines, combinations, discoveries, formulae, manufacturing techniques, technical developments, artwork, software, programming, applets, scripts, designs, or other business processes.

Intellectual Property Rights: All patent, patent rights, patent applications, copyrights, copyright registrations, trade secrets, trademarks, and service marks.

Interface: A means of allowing two or more elements of software, hardware, or other systems to exchange data and otherwise "talk" to each other.

Internet: A global collection of interconnected computers and networks that use the TCP/IP (Transmission Control Protocol/Internet Protocol) protocols to communicate with each other. At one time, the term "Internet" was used as an acronym for "interconnected networks."

Intranet: A computer network designed to be used within a business or company. An intranet is so named because it uses much of the same technology as the Internet. Web browsers, e-mail, newsgroups, HTML documents, and websites are all found on intranets. In addition, the method for transmitting information on these networks is TCP/IP. See also **Internet**.

ISP: Acronym for Internet Service Provider.

ITIL: The Information Technology Infrastructure Library.

Joint Patent: Patents that are jointly owned by the parties.

JPG: Acronym for Joint Photographic Experts Group. A standard file format for storing images in a compressed form.

LAN: Acronym for local area network. See **Local Area Network**.

Local Area Network: Abbreviated "LAN." A network of computers and other devices generally located within a relatively limited area (for example, within a particular office, building, or group of buildings).

Log File: A record of activity or transactions that occur on a particular computer system.

Losses or Loss: Any and all third-party claims, liabilities, losses, costs, and damages, including reasonable attorneys' fees, expert witness fees, expenses, and costs of settlement paid by, or that become payable by, the indemnified party to the applicable third-party claimant arising out of or with respect to the agreement.

Managed Strategic Provider: Third-party vendors identified by client as having responsibility under a separate agreement with client for the delivery of a critical service segment that must integrate with the services.

Metadata: Information maintained on a computer system relating to an electronic document, but that is not generally printed when the document is reduced to hard copy form. Metadata may include, but is not limited to, information about the person who created the document, the data and time of creation, the number of times the document was edited and by whom, and the program used to create the document.

Minor Enhancements: Discrete requests for the introduction of changes that: (i) modify or add functionality to any then-existing application or applications (e.g., changes to inputs, data, results, reports, or processing logic); (ii) that are reviewed, approved, and funded by client; (iii) that require three-hundred twenty (320) hours or less of provider resources as set forth in the applicable work order or other approval at the time the work is approved; and (iv) are subject to the change management procedures. Provider activities to perform a minor enhancement exclude provider activities required for production support and corrective maintenance as defined in the agreement.

MIS: Management Information Services. The department within a business assigned the task of operating and maintaining its computer resources.

Modem: A device used to change a stream of data from a computer into audio tones for transmission over a telephone line. Originally a concatenation of Modulator-Demodulator.

Moral Rights: Any personal or noneconomic right to a work, including rights of attribution, integrity of the work, any right to object to any distribution or other modification of a work, and any similar right existing under the law of any country in the world or under any treaty.

MPEG: Acronym for Moving Pictures Experts Group. Standard file format for storing compressed versions of audio and video files.

NDA: An acronym for Nondisclosure Agreement. See **Nondisclosure Agreement**.

Network Map: A network map is a graphical depiction of the way in which the various computers, file servers, and peripherals on a network are interconnected. The map typically identifies the type and speed (bandwidth) of the connections.

New Services: Those services that are different in purpose from, and in addition to, the Services. All new services should require client's approval.

Nonbillable Project: Services provided pursuant to client or, as applicable, provider's, project management methodology then in place.

Nondisclosure Agreement (NDA): An agreement, generally entered into at an early stage in a potential engagement, that governs the parties' respective confidentiality obligations.

NOS: Acronym for network operating system. See also **File Server**.

Object Code: The machine readable version of a computer program. See **Source Code**.

OCC: An acronym for the Office of the Comptroller of the Currency, a regulator in the financial services industry.

Offshore: In the context of a professional service engagement, contractors who are located outside the United States.

Open Source Software: Any Intellectual Property that is subject to the GNU General Public License, GNU Library General Public License, Artistic License, BSD License, Mozilla Public License, or any similar license, including, those licenses listed at www.opensource.org/licenses.

Operating System: Abbreviated OS. A program used to control the basic operation of a computer (for example, storing and retrieving data from memory, controlling how information is displayed on the computer monitor, operating the central processing unit, and communicating with peripherals).

PC: Acronym for personal computer.

Partition: A region of a hard disk treated by the computer's operating system as a separate drive. Through the use of partitions, a computer with a single hard disk can appear to have two or more drives.

PCMCIA: Acronym for Personal Computer Memory Card International Association. A standard for laptop and PDA-based peripherals (for example, memory cards and modems) and the slots that hold them. See also **Personal Digital Assistants**.

Peer-to-Peer Network: A type of network in which a group of personal computers is interconnected so that the hard disks, CD ROMS, files, and printers of each computer can be accessed from every other computer on the network. Peer-to-peer networks do not have a central file server. This type of system is used in situations where less than a dozen computers will be networked.

Performance Credits: A credit to which client becomes entitled pursuant to the agreement. All performance credits described are to reflect in part the diminished value for the services delivered as compared to the service levels, or other contractual commitments, and, unless otherwise set forth in the agreement, do not represent all damages, penalties, or other compensation remedy that may result from any failure to meet such service levels or other contractual requirements.

Peripherals: A device, such as a printer, mouse, or disk drive, that is connected to a computer and controlled by its microprocessor.

Personal Digital Assistants: Abbreviated "PDA." PDA's range from compact personal electronic organizers (for example, calendars, phone lists, brief notes) to the new breed of palm-sized computers that are capable of running full-featured word-processing programs and spreadsheets and of browsing the Internet and sending and receiving e-mail. These devices can hold hundreds, and soon thousands, of pages of information.

PII: Acronym for Personally Identifiable Information.

Platform as a Service (PAAS): The capability provided to the customer to deploy onto the cloud infrastructure customer-created or -acquired applications created using programming languages and tools supported by the provider. The consumer does not manage or control the underlying cloud infrastructure, including network, servers, operating systems, or storage, but has control over the deployed applications and possibly application hosting environment configurations.

Privity: Refers to instances in which two entities are both parties to the same agreement, resulting in "contractual privity." In this way, each party can directly enforce the contract against the other party.

Problem: An unknown underlying cause of one or more Incidents.

Problem and Incident Management System: An integrated and automated system that is used by provider to manage the process of monitoring, identifying, logging, tracking, escalating, reviewing, resolving and reporting problems, incidents, changes, and other issues related to the services and the system. The problem and incident management system will integrate with the appropriate software, equipment, electronic mail system, and provider's knowledge databases. As of the effective date, provider will use remedy as the problem and incident management system.

Provider Work: Tangible and intangible information and developments that are owned by provider, including all intermediate and partial versions thereof and all designs, specifications, materials, program materials, software, flowcharts, outlines, lists, compilations, manuscripts, writings, pictorial materials, schematics, other creations, and the like.

Proxy Server: A server used to manage Internet-related traffic coming to and from a local area network and can provide certain functionality (for example, access control and caching of popular websites).

Provider's Service Delivery Manager or Provider's SDM: Provider's primary person responsible for the overall quality of service delivery to client and for ensuring that superior levels of performance and customer service are achieved.

Provider's SDM will lead provider's operational team responsible for delivering managed services and will have responsibility for managing the technical resources performing services on the client's account and for ensuring that their tasks are prioritized according to client's requirements. Provider's SDM will be located onsite at Client's primary Service Location in Natick, Massachusetts, on a full-time basis. The client technical and management teams will have continuous access to provider's SDM to ensure that changing priorities, needs, and support requirements are quickly communicated to the provider technical resources for timely and appropriate response.

Public Key Cryptography: An encryption method that uses a two-part key: a public key and a private key. Users generally distribute their public key, but keep their private key to themselves. This is also known as Asymmetric Cryptography.

RAM: Acronym for random access memory. See **Random Access Memory**.

Ramp-up Time: Use in the context of professional service engagements to refer to the time required to bring a new worker up-to-speed on a project and become productive.

Random Access Memory: Abbreviated RAM. An integrated circuit into which data can be read or written by a microprocessor or other device. The memory is volatile and will be lost if the system is disconnected from its power source.

Rate Card: In professional service engagements, this is a contract exhibit that lists the various hourly rates charged by the contractor's personnel.

Read-only Memory: Abbreviated ROM. An integrated circuit into which information, data, and/or programs are permanently stored. The absence of electric current will not result in loss of memory.

Release: A redistribution of Software that contains new features, new functionality, or performance improvements.

Release Condition: The conditions specified in a Source Code Escrow Agreement under which the escrow agent is required to release the source code. Typical release conditions include insolvency, bankruptcy, and abandoning support obligations.

Request for Information (RFI): A means to gather information from suppliers and vendors about their products and services. An RFI may be conducted through a formal written document (something similar to an RFP) or a letter outlining general project concepts. It may also be conducted more informally through personal communication.

Request for Proposal (RFP): An invitation for suppliers and vendors, generally through a competitive bidding process, to submit a proposal to furnish a specific commodity or service.

Reseller: A company that is in the business of reselling products and services created by others.

Resolve or Resolution: To correct an Incident or Problem with either a permanent solution or an interim workaround solution (including fixes, patches, upgrades, defect repairs, program and corrections, and remedial programming); provided that, when an interim workaround solution is used, Provider shall use its best efforts to continuously address the underlying issue, incident, or problem to promptly provide a permanent resolution. In addition to the description in the foregoing sentence, where a ticket is generated for an issue, incident, or problem, "resolve" or "resolution" shall mean the end-to-end conclusion of all ticket resolution efforts (by all support levels) on a ticket's underlying issue, incident, or problem. Ticket closure is not complete and a ticket may not be closed unless the authorized user confirms that the underlying issue, incident, or problem has been resolved. In the event that an authorized user determines an issue to still be unresolved after a ticket is marked as resolved, the ticket will be changed back to open status and the total elapsed time to resolve the issue will be continued for that ticket from the time the ticket was put into resolved status, as if the ticket had not been closed then re-opened. For example, if (i) a ticket was put in resolved status after three (3) hours of work by provider, (ii) an authorized user determines an issue to still be unresolved after such ticket was put in resolved status, (iii) the ticket is reopened, (iv) the ticket is then put in resolved status again after an additional three (3) hours of work by provider, and (v) the authorized user was then satisfied that all issues are resolved and does not, subsequently, seek to have the ticket re-opened, then (vi) the total time spent on that particular ticket is six (6) hours, regardless of the fact that the ticket was put in resolved status multiple times. Unless otherwise defined in the applicable statement of work, ticket resolution efforts may involve either a permanent solution or an interim workaround solution (including fixes, patches, upgrades, defect repairs, programming corrections, and remedial programming); provided that, when an interim workaround solution is used, provider shall use its best efforts to continuously address the underlying issue, incident, or problem to promptly provide a permanent resolution.

RFI: An acronym for Request for Information. See **Request for Information**.

RFP: An acronym for Request for Proposal. See **Request for Proposal**.

ROM: Acronym for read only memory. See **Read-only Memory**.

Root Cause Analysis: An analysis performed by Provider in order to determine the reason for Provider's failure to meet its obligations under the Agreement.

Sarbanes–Oxley Act: The Public Company Accounting Reform and Investor Protection Act of 2002 and its implementing regulations.

Scanner: A piece of computer hardware used to scan text and/or images into the memory of a computer system.

Scheduled Uptime: The period of time during which Services are to be available to all Authorized Users for normal business use, expressed in hours and minutes.

Scoping Statement of Work: An initial statement of work that is frequently used to afford the parties the opportunity to explore the effort needed to complete a larger project and define its constraints. The output of the scoping statement of work provides the input for drafting a further statement of work for the larger project contemplated by the parties.

Self-Escrow: A source code escrow arrangement in which the customer (rather than a third-party escrow agent) is given possession of the source code, but must maintain the source code unused in a secure location (e.g., a safe) until a release condition occurs.

Server: Any computer that provides shared processing or resources (e.g., printer, fax, application processing, database, mail, proxy, firewalls, and backup capabilities) to authorized users or other computers over the system. A Server includes associated peripherals (e.g., local storage devices, attachments to centralized storage, monitor, keyboard, pointing devices, tape drivers, and external disk arrays) and is identified by a unique manufacturer's serial number.

Service Agreement: A legal document specifying the rules of the legal contract between the cloud user and the cloud provider.

Service-Level Agreement: A document stating the technical performance promises made by the cloud provider, how disputes are to be discovered and handled, and any remedies for performance failures.

Service Levels: Provider performance criteria as set forth in the service level agreement and shall include critical service levels and key performance indicators.

Service Request: The form by which client submits requests to provider for changes, additions, modifications, or enhancements to the services and software that are not technological improvements as defined herein.

Services: All of the tasks and services described in the agreement, and all of the tasks and services that are described in the exhibits and schedules attached to the agreement, and any tasks, functions, services, responsibilities, and obligations not specifically described in the agreement, but that are inherent subtasks of such tasks, functions, services, responsibilities, and obligations and are incidental or necessary to, reasonably inferred to be within, and are required for the proper performance and provision of the services described in the agreement and in the exhibits and schedules attached to the agreement, other than those identified in the agreement as tasks and services to be performed by client.

Severity: The priority assigned to a reported ticket (including any issue, problem, incident, or service request) with respect to any service. The process for categorization of tickets into Severity 1, Severity 2, or Severity 3 will consider the impact on service availability and performance, availability of a workaround, and the number of people impacted. In the event provider assigns a severity status designation to a ticket that client perceives to be incorrect, then client shall have the right to dispute and change the severity status designation on an expedited basis.

Severity 1: An emergency or critical issue requiring immediate attention and resolution. An issue, service request, incident, or problem shall be classified as Severity 1 if: (i) a major impact to multiple components in the computing environment; (ii) a mission critical component is down; (ii) a service level is being missed; or (iv) a key business condition documented or identified by service management or the client is impacted. Resolution efforts for Severity 1 issues, service requests, incidents, or problems will be initiated immediately and attention to the resolution of these issues, service requests, incidents, or problems will be sustained around the clock until a temporary workaround is in place or a permanent solution is implemented.

Severity 2: Issues, tickets, service requests, incidents, or problems with high impact to client's business operations. Rapid, not necessarily immediate, restoration required. An issue, tickets, service request, incident, or problem shall be classified as Severity 2 if: (i) there's a critical application or software problem; (ii) the system or services are degraded for multiple end users; (iii) a key component of the system or services is severely degraded; (iv) a key component of the system or services is in danger of failing; or (v) a service level is degraded or in danger of being missed. Resolution efforts for Severity 2 issues, tickets, service requests, incidents, or problems will be initiated immediately and attention to the resolution of these two issues, tickets, service requests, incidents, or problems will be sustained around the clock until a temporary workaround is in place or a permanent solution is implemented.

Severity 3: Issues, tickets, service requests, incidents, or problems that have a medium impact to client. An issue, tickets, service request, incident, or problem shall be classified as Severity 3 if: (i) a noncritical component of the system or services is in danger of failing; **(ii) minor components of the system or services are down or failing; and (iii) service or performance is degraded for a single user. Severity 4:** Issues, tickets, service requests, incidents, or problems that have a low impact to Client. An issue, tickets, service request, incident, or problem shall be classified as Severity 4 if: (i) the system or services have not been impacted; (ii) an alternative method is being used with no impact or service degradation to the system or services; and (iii) there is no business impact to the System or Services.

Shrink-Wrap Agreement: An agreement that is included as part of the packaging or in the documentation accompanying a piece of software or equipment. In some cases, the CD containing the software may be provided in an envelope with the shrink-wrap agreement printed on the outside. Opening of the envelope indicated the user's acceptance of the terms.

Software: A series of prerecorded commands issued to a computer to accomplish a particular task. Individually each, and collectively all, of the system software, applications, computer programs and other software, licensed by client or provider from a third-party vendor, or otherwise provided by provider or client under the agreement, including the problem and incident management system and any: (i) embedded or re-marketed third-party vendor software or computer programs; (ii) interfaces; (iii) source materials; or (iv) object code. Software shall include any and all revisions thereto, and any and all programs provided by a third-party vendor, provider, or client in the future under the agreement.

Software as a Service (SaaS): The capability provided to the consumer to use the provider's applications running on a cloud infrastructure. The applications are accessible from various client devices through a thin client interface such as a web browser (e.g., webbased e-mail). The consumer does not manage or control the underlying cloud infrastructure including network, servers, operating systems, storage, or even individual application capabilities, with the possible exception of limited user-specific application configuration settings.

Software Developers Kit or SDK: A type of license agreement under which a licensor provides the licensee with certain information, most notably interface specifications, to allow the licensee to create software that interacts with systems and software of the licensor.

Source Code: The version of a computer program that can be read by humans. The source code is translated into machine readable code by a program called a "compiler." Access to the source code is required to understand how a computer program works or to modify the program. See **Object Code**.

Source Code Escrow Agreement: An agreement between a software licensor and a third-party escrow company to hold in trust the source code for certain of the licensor's software. On the occurrence of certain defined conditions, the escrow company will release the source code to the beneficiary of the escrow (usually the licensee of the object code version of the software).

Source Materials: With respect to software, the source code of such software and all related compiler command files, build scripts, scripts relating to the operation and maintenance of such software, application programming interface (API), graphical user interface (GUI), object libraries, all relevant instructions on building the object code of such software, and all documentation relating to the foregoing.

Spaghetti Code: Source code that was not developed using accepted programming practices and that is poorly written and documented.

Stand-alone Computer: A personal computer that is not connected to any other computer or network, except possibly through a modem.

Statement of Work (SOW): A written description of the services, milestones, deliverables, project schedule, and fees associated with professional services to be rendered by a contractor or vendor. The statement of work is generally attached to an overarching professional services or similar contract.

Submission: Written or verbal suggestions, comments, ideas, feedback, etc., from a third party to a business about the business' products and services.

Submission Agreement: See **Idea Submission Agreements**.

System: The functions, capabilities, operations, equipment, computer programs, other resources (other than human resources), and systems that, at any particular time, are used, operated, supported, or maintained by provider, on behalf of, or for the benefit of, client in provider's performance of the services hereunder, including as components thereof: (i) the software and equipment; (ii) the work product; (iii) the documentation; and (iv) client system and the entire system of hardware, software, equipment, environments, networks, and network components that constitute, are associated with or related to, or interconnect, any of the items described by the foregoing clauses (i) through (iv), or on which such items are installed, operated, or used, including any such hardware and software produced by third parties embedded within such software or the work product, and revisions, updates, modifications and customizations to any or all of client system and the entire system of the hardware, software, documentation, equipment, networks and network components described by such clauses, in accordance with the agreement.

System Software: Those programs and programming (including the supporting documentation, media, online help facilities, and tutorials) that perform tasks basic to the functioning of client equipment and which are required to support the operations of the applications or otherwise support the provision of services by provider. System software includes server operating systems, network operating systems, systems utilities (including measuring and monitoring tools), data security software, middleware, and telecommunications monitors.

Technological Improvements: Any supplementation, modification, enhancement, and replacement of the technology employed by provider in the provision of the services in the normal course of business during the term.

Term Sheet: A document containing the key terms of a proposed transaction. The term sheet is generally exchanged between the supplier and the customers to define the outline of proposed key terms for a transaction (e.g., pricing, quantity of purchase, timing, etc.).

Third Party: Means an entity that is not in contractual privity (e.g., a typical vendor subcontractor is not a party to the agreement between the vendor and the customer). In these situations, it is not possible to directly enforce the contract against the third party.

Third-Party Vendor: Any person or entity (excluding client or provider) contracting directly or indirectly with client or provider to provide equipment, intellectual property, services or other products or services that are used or provided under the agreement.

Third-Party Works: Tangible and intangible information and developments that are owned by neither Party, including all intermediate and partial versions thereof and all designs, specifications, materials, program materials, software, flow charts, outlines, lists, compilations, manuscripts, writings, pictorial materials, schematics, other creations, and the like.

Three-Party Escrow Agreement: A source code escrow agreement signed by the software licensor, escrow agent, and the licensee "Verification Services" means the services offered by an escrow agent to verify that the deposited source code matches the version of the software currently being licensed by the vendor.

Ticket: A single electronic record of an Authorized User request for service initiated through help desk, a web-enabled interface or other means, or a Service Request for onsite support that is created to document the steps taken to resolve such Authorized User request.

Tier 1 Efforts: Those efforts associated with the highest echelon of application service providers and business process outsourcers to the retail industry. The parties acknowledge and agree that the term Tier 1 Efforts is intended to mean a level of effort higher than commercially reasonable efforts and lower than best efforts.

Tier 1 Support: The provision of help desk support provided by client or provider.

Tier 2 Support: The next level of support for incidents assigned to provider by the client in accordance with the authorization matrix. If Tier 1 Support does not resolve the incident or question, the incident shall be referred to Tier 2 Support. Tier 2 support provides a higher level of application expertise and knowledge that goes beyond Tier 1 Support.

Tier 3 Support: The provision of advanced technical support, consultation and guidance to Tier 1 Support and Tier 2 Support team by provider subject matter experts and the assistance in the final resolution of complex incidents. Tier 3 Support also involves interacting with other development teams or third-party vendors providing application support that provide consultation in incident determination and resolution, including the development of programming workarounds, patches, and fixes. This support provides specific system or product knowledge or programming expertise.

Time and Materials (T&M): A professional service engagement in which the contractor will be paid an hourly or daily rate, plus expenses.

Tools: Any software, utilities, know-how, methodologies, processes, technologies or algorithms used by provider in providing the services and monitoring the service levels.

Two-Party Escrow Agreement: A source code escrow agreement that is signed by the software licensor and the escrow agent. The licensee is not a party to the agreement, but signs a beneficiary form or other document entitling the licensee to receive a copy of the source code on the occurrence of a release condition.

T&M: Acronym for time and materials. See **Time and Materials (T&M)**.

VAN: Acronym for value added network. See **Value-Added Network**.

Value-Added network: Abbreviated "VAN." A computer communications network commonly used by businesses to exchange EDI documents.

Virtual Machine (VM): An efficient, isolated duplicate of a real machine.

Virtual Memory: The ability of a computer operating system to make RAM appear larger than it really is by storing programs and data temporarily on the hard disk. See **Swap Files**.

Virtualization: The simulation of the software and/or hardware upon which other software runs.

WAN: Acronym for wide area network. See **Wide Area Network**.

Web: See **World Wide Web**.

Web Browser: A program used to view HTML pages on the World Wide Web.

Web Server: A computer on which a website is stored.

Web-Wrap Agreement: A click-wrap agreement or other form of terms and conditions presented to the user in connection with use of a website or online service. The standard terms and conditions of use commonly found as a hyperlink on the first page of a website is an example of a web-wrap agreement.

Website: A collection of related HTML documents stored on the same computer and accessible to users of the Internet. See also **Home Page**.

Wide Area Network: Abbreviated "WAN." A network of computers and other devices distributed over a broad geographic area.

Work Product: All information and developments and all intermediate and partial versions thereof, including all source code and object code with respect thereto, all applications, scripts, processes, methods, and programs related to either responding to, processing, handling, tracking, or fulfilling client's requests for information and assistance and all related policies and procedures, and all designs, specifications, improvements, materials, program materials, software, flow charts, outlines, lists, compilations, manuscripts, writings, pictorial materials, schematics, other creations, and the like, whether or not subject to copyright or otherwise able to be protected by law, created, invented, or conceived for client's use or benefit in connection with the agreement, by client, by provider, or by any other person engaged by client or provider. All work product created by provider constitute deliverables. For the avoidance of doubt, work product shall not include any portions or derivatives of provider work.

Workstation: A personal computer connected to a network. A workstation can also refer to a high performance computer used for intensive graphics or numerical calculations.

World Wide Web: Abbreviated WWW or the web. A user-friendly, graphical interface to the Internet. Documents, sometimes called "pages," on the web are created using the hypertext markup language (HTML). Documents are connected to one another on the web through the use of "hyperlinks."

A hyperlink is a highlighted word, phrase, or graphic image that when selected by pressing a key or the click of a mouse will automatically transfer the user from one page on the web to another. Web pages are transmitted over the Internet using the hypertext transfer protocol (HTTP).

WWW: Acronym for World Wide Web. See **World Wide Web**.

Federal Financial Institutions Examination Council

FFIEC

Outsourcing Technology Services

JUNE 2004

IT EXAMINATION HANDBOOK

TABLE OF CONTENTS

INTRODUCTION

The financial services industry has changed rapidly and dramatically. Advances in technology enable institutions to provide customers with an array of products, services, and delivery channels. One result of these changes is that financial institutions increasingly rely on external service providers for a variety of technology-related services. Generally, the term "outsourcing" is used to describe these types of arrangements.

The Federal Financial Institutions Examination Council (FFIEC) *Information Technology Examination Handbook* (IT Handbook) "Outsourcing Technology Services Booklet" (booklet) provides guidance and examination procedures to assist examiners and bankers in evaluating a financial institution's risk management processes to establish, manage, and monitor IT outsourcing relationships.

The ability to contract for technology services typically enables an institution to offer its customers enhanced services without the various expenses involved in owning the required technology or maintaining the human capital required to deploy and operate it. In many situations, outsourcing offers the institution a cost effective alternative to in house capabilities. Outsourcing, however, does not reduce the fundamental risks associated with information technology or the business lines that use it. Risks such as loss of funds, loss of competitive advantage, damaged reputation, improper disclosure of information, and regulatory action remain. Because the functions are performed by an organization outside the financial institution, the risks may be realized in a different manner than if the functions were inside the financial institution resulting in the need for controls designed to monitor such risks.

Financial institutions can outsource many areas of operations, including all or part of any service, process, or system operation. Examples of information technology (IT) operations frequently outsourced by institutions and addressed in this booklet include: the origination, processing, and settlement of payments and financial transactions; information processing related to customer account creation and maintenance; as well as other information and transaction processing activities that support critical banking functions, such as loan processing, deposit processing, fiduciary and trading activities; security monitoring and testing; system development and maintenance; network operations; help desk operations; and call centers. The booklet addresses an institution's responsibility to manage the risks associated with these outsourced IT services.

Management may choose to outsource operations for various reasons. These include

- Gain operational or financial efficiencies;
- Increase management focus on core business functions;
- Refocus limited internal resources on core functions; Obtain specialized expertise;
- Increase availability of services;

- Accelerate delivery of products or services through new delivery channels;
- Increase ability to acquire and support current technology and avoid obsolescence; and
- Conserve capital for other business ventures.

Outsourcing of technology-related services may improve quality, reduce costs, strengthen controls, and achieve any of the objectives listed previously. Ultimately, the decision to outsource should fit into the institution's overall strategic plan and corporate objectives.

Before considering the outsourcing of significant functions, an institution's directors and senior management should ensure such actions are consistent with their strategic plans and should evaluate proposals against well-developed acceptance criteria. The degree of oversight and review of outsourced activities will depend on the criticality of the service, process, or system to the institution's operation.

Financial institutions should have a comprehensive outsourcing risk management process to govern their technology service provider (TSP) relationships. The process should include risk assessment, selection of service providers, contract review, and monitoring of service providers. Outsourced relationships should be subject to the same risk management, security, privacy, and other policies that would be expected if the financial institution were conducting the activities in-house. This booklet primarily focuses on how the bank regulatory agencies review the risk management process employed by a financial institution when considering or executing an outsourcing relationship.

To help ensure financial institutions operate in a safe and sound manner, the services performed by TSPs are subject to regulation and examination.[1] The federal financial regulators have the statutory authority to supervise all of the activities and records of the financial institution whether performed or maintained by the institution or by a third party on or off of the premises of the financial institution. Accordingly, the examination and supervision of a financial institution should not be hindered by a transfer of the institution's records to another organization or by having another organization carry out all or part of the financial institution's functions.[2]

Many of the general principles on effective management of outsourcing relationships discussed in this booklet can and should be applied to managing the outsourcing of software development. Outsourcing of activities related to software development is addressed in the *IT Handbook's*, "Development and Acquisition Booklet."

This booklet rescinds and replaces Chapter 22 of the *1996 FFIEC Information Systems Examination Handbook, IS Servicing – Provider and Receiver.*

---

[1] See 12 USC 1867 (c)(1) and 12 USC 1464 (d)(7). The NCUA does not currently have independent regulatory authority over TSPs.

[2] S. Rep. No. 2105, 87-2105 at 3 (1962). Reprinted in 1962 U.S.C.C.A.N. 3878, 3880. Accord H.R. Rep. No. 105-417, at 4 (1998), reprinted in 1998 U.S.C.C.A.N. 22. 23.

BOARD AND MANAGEMENT RESPONSIBILITIES

Action Summary

The financial institution's board and senior management should establish and approve risk-based policies to govern the outsourcing process. The policies should recognize the risk to the institution from outsourcing relationships and should be appropriate to the size and complexity of the institution.

The responsibility for properly overseeing outsourced relationships lies with the institution's board of directors and senior management. Although the technology needed to support business objectives is often a critical factor in deciding to outsource, managing such relationships is more than just a technology issue; it is an enterprise-wide corporate management issue. An effective outsourcing oversight program should provide the framework for management to identify, measure, monitor, and control the risks associated with outsourcing. The board and senior management should develop and implement enterprise-wide policies to govern the outsourcing process consistently. These policies should address outsourced relationships from an end-to-end perspective, including establishing servicing requirements and strategies; selecting a provider; negotiating the contract; and monitoring, changing, and discontinuing the outsourced relationship.

Factors institutions should consider include:

- Ensuring each outsourcing relationship supports the institution's overall requirements and strategic plans;
- Ensuring the institution has sufficient expertise to oversee and manage the relationship;
- Evaluating prospective providers based on the scope and criticality of outsourced services;
- Tailoring the enterprise-wide, service provider monitoring program based on initial and ongoing risk assessments of outsourced services; and
- Notifying its primary regulator regarding outsourced relationships, when required by that regulator.[1]

The time and resources devoted to managing outsourcing relationships should be based on the risk the relationship presents to the institution. To illustrate, outsourcing processing of a small credit card portfolio will require a different level of

[1] 12 USC 1867 (c) (11), Bank Service Company Act (Banks), and 12 USC 1464 (d) (7) Home Owners' Loan Act (Thrifts). In addition, Thrift Bulletin 82, Third Party Arrangements (March 18, 2003), states that thrifts should notify the Office of Thrift Supervision at least 30 days before establishing a third party relationship with a foreign service provider.

oversight than outsourcing processing of all loan applications. Additionally, smaller and less complex institutions may have less flexibility than larger institutions in negotiating for services that meet their specific needs and in monitoring their service providers.

RISK MANAGEMENT

Risk management is the process of identifying, measuring, monitoring, and managing risk. Risk exists whether the institution maintains information and technology services internally or elects to outsource them. Regardless of which alternative they choose, management is responsible for managing risk in all outsourcing relationships. Accordingly, institutions should establish and maintain an effective risk management process for initiating and overseeing all outsourced operations.

An effective risk management process involves several key factors:

- Establishing senior management and board awareness of the risks associated with outsourcing agreements in order to ensure effective risk management practices;
- Ensuring that an outsourcing arrangement is prudent from a risk perspective and consistent with the business objectives of the institution;
- Systematically assessing needs while establishing risk-based requirements;
- Implementing effective controls to address identified risks;
- Performing ongoing monitoring to identify and evaluate changes in risk from the initial assessment; and
- Documenting procedures, roles/responsibilities, and reporting mechanisms.

Typically, this process incorporates the following activities:

- Risk assessment and requirements definition;
- Due diligence in selecting a service provider;
- Contract negotiation and implementation; and
- Ongoing monitoring.

The preceding comments focus on risk elements specifically associated with outsourcing. For a broader perspective on IT transactional and operational risk, refer to the IT Handbook's "Supervision of Technology Service Providers (TSP) Booklet," which addresses outsourcing risk from the service provider perspective.

RISK ASSESSMENT AND REQUIREMENTS

Action Summary

Management should:

- Assess the risk from outsourcing;
- Involve stakeholders in creating risk-based written requirements to control an outsourcing action; and
- Use the written requirements to guide and manage the remainder of the outsourcing process.

Outsourced IT services can contribute to operational risks (also referred to as transaction risks). Operational risk may arise from fraud, error, or the inability to deliver products or services, maintain a competitive position, or manage information. It exists in each process involved in the delivery of the financial institutions' products or services. Operational risk not only includes operations and transaction processing, but also areas such as customer service, systems development and support, internal control processes, and capacity and contingency planning. Operational risk also may affect other risks such as interest rate, compliance, liquidity, price, strategic, or reputation risk as described below.

- *Reputation risk*—Errors, delays, or omissions in information technology that become public knowledge or directly affect customers can significantly affect the reputation of the serviced financial institutions. For example, a TSP's failure to maintain adequate business resumption plans and facilities for key processes may impair the ability of serviced financial institutions to provide critical services to their customers.
- *Strategic risk*—Inadequate management experience and expertise can lead to a lack of understanding and control of key risks. Additionally, inaccurate information from TSPs can cause the management of serviced financial institutions to make poor strategic decisions.
- *Compliance (legal) risk*—Outsourced activities that fail to comply with legal or regulatory requirements can subject the institution to legal sanctions. For example, inaccurate or untimely consumer compliance disclosures or unauthorized disclosure of confidential customer information could expose the institution to civil money penalties or litigation. TSPs often agree to comply with banking regulations, but their failure to track regulatory changes could increase compliance risk for their serviced financial institutions.
- *Interest rate, liquidity, and price (market) risk*—Processing errors related to investment income or repayment assumptions could lead to unwise investment or liquidity decisions thereby increasing market risks.

QUANTITY OF RISK CONSIDERATIONS

The quantity of risk associated with an outsourced IT service is subject to the function outsourced, the service provider, and the technology used by the service provider. Management should consider the following factors in evaluating the quantity of risk at the inception of an outsourcing decision.

- Risks pertaining to the function outsourced include:
  - Sensitivity of data accessed, protected, or controlled by the service provider;
  - Volume of transactions; and
  - Criticality to the financial institution's business.
- Risks pertaining to the service provider include:
  - Strength of financial condition;
  - Turnover of management and employees; Ability to maintain business continuity;
  - Ability to provide accurate, relevant, and timely Management Information Systems (MIS);
  - Experience with the function outsourced;
  - Reliance on subcontractors;
  - Location, particularly if cross-border (See Appendix C, Foreign-Based Third-Party Service Providers); and
  - Redundancy and reliability of communication lines.
- Risks pertaining to the technology used include:
  - Reliability;
  - Security; and
  - Scalability to accommodate growth.

REQUIREMENTS DEFINITION

The definition of business requirements sets the stage for all outsourcing actions and forms the basis for subsequent management of the outsourced activity. The requirements are developed through a process that identifies the functions or activities to be outsourced, assesses the risk of outsourcing those functions or activities, and establishes a baseline from which appropriate control measures can be identified. These requirements provide a basis for an understanding between the financial institution and the service provider as to what the risks are and how they will be managed and controlled.

Key Practices

Sound practices for the development of requirements include:

- *Stakeholder involvement*—All organizational groups who will be directly involved with the service provider or in using the contracted service should be represented in the development of product and service requirements.

- *Integration*—The development should result in requirements that support the subsequent steps of solicitation, selection, contracting, and monitoring.
- *Documentation*—Documentation will greatly assist in ensuring that the service contracted and delivered meets the institution's requirements. Documentation will also allow for subsequent reviews of the processes' adequacy and integrity.

Components

The requirements definition phase should result in a detailed document containing descriptions of the institution's expectations relative to the outsourced service. The requirements document may consider, but is not limited by, the following high level topical components:

Scope and nature

- Service description;
- Technology; and
- Customer support.

Standards and service levels

- Availability and performance;
- Change management;
- Financial reporting;
- Quality of service;
- Security; and
- Business continuity.

Minimum acceptable service provider characteristics

- Industry experience;
- Management experience;
- Technology and systems architecture;
- Process controls;
- Financial condition;
- Reputation, including references;
- Degree of reliance on third parties, subcontractors, or partners;
- Legal, regulatory, and compliance history; and
- Ability to meet future needs.

Monitoring and reporting

- Measurements and reporting criteria;
- Right to audit;
- Third-party reports; and
- Coordination of responses to security events.

Transition requirements

- Initial migration of data to the service provider;
- Implementation of necessary communications mechanisms;

- Migration of data from the service provider at termination of contract; and
- Staff training.

Contract duration, termination, and assignment

- Start and term;
- Conditions and right to cancel;
- Ownership of data;
- Timely return of data in machine-readable format;
- Costs of transition;
- Limitations, as appropriate, governing assignment to third party;
- Dispute resolution; and
- Confidentiality of institution data.

Contractual protections against liability

- Indemnification;
- Limitation of liability; and
- Insurance.

When outsourcing to a subsidiary or affiliate is considered, management must assure that the components outlined above evidence an arms-length transaction. An arrangement between a financial institution and an affiliate or subsidiary should be on terms that are substantially the same, or at least as favorable to the institution, as those prevailing at the time for comparable transactions with a non-affiliated third party.

SERVICE PROVIDER SELECTION

Action Summary

Management should:

- Evaluate service provider proposals in light of the institution's needs, including any differences between the institution's solicitation and the service provider proposal;
- Perform due diligence on the prospective service providers;
- Ensure that selection of affiliated parties as service providers is done at arms length in accordance with regulations and guidance issued by the institution's primary regulator; and
- Evaluate foreign-based third-party service providers in light of the guidance found in this section and in Appendix C, Foreign-Based Third-Party Service Providers.

After identifying the work to be performed and the necessary controls, a financial institution solicits responses from prospective service providers. The primary tool for the solicitation is the Request for Proposal (RFP). The RFP also supports subsequent contract negotiations.

REQUEST FOR PROPOSAL

A financial institution should generate the RFP from the information developed during the requirements definition phase. While the level of detail may vary for any particular procurement, the RFP should describe the institution's objectives; the scope and nature of the work to be performed; the expected production service levels, delivery timelines, measurement requirements, and control measures; and the financial institution's policies for security, business continuity, and change control. It also requests responses addressing those requirements as well as the fees each service provider will charge.

Once management distributes the RFPs and receives responses, it should evaluate the service provider proposals against the institution's needs. When the institution evaluates the proposals, it may find that the proposals do not completely agree with the RFP. For example, the service the service provider proposes may include different processing workflows or reporting schemes, pricing formulas or techniques, or the response to information requests may not be complete. If the institution considers proposals that differ from the RFP, the institution should evaluate the differences against its requirements and clearly understand how the changes will affect the institution's objectives and service expectations.

The institution should evaluate material differences using a process similar to the one used to develop the requirements initially. An institution should negotiate a resolution to any differences between the RFP and the service provider proposal before contracting with a service provider.

DUE DILIGENCE

A financial institution should perform due diligence on the service provider's response to an RFP as well as the service provider itself. Due diligence should serve as a verification and analysis tool, providing assurance that the service provider meets the institution's needs. Due diligence should confirm and assess the following information regarding the service provider:

- Existence and corporate history;
- Qualifications, backgrounds, and reputations of company principals, including criminal background checks where appropriate;
- Other companies using similar services from the provider that may be contacted for reference;
- Financial status, including reviews of audited financial statements;
- Strategy and reputation;
- Service delivery capability, status, and effectiveness;
- Technology and systems architecture;
- Internal controls environment, security history, and audit coverage;
- Legal and regulatory compliance including any complaints, litigation, or regulatory actions;
- Reliance on and success in dealing with third party service providers;
- Insurance coverage; and
- Ability to meet disaster recovery and business continuity requirements.

Other important elements include probing for information on intangibles, such as the third party's service philosophies, quality initiatives, and management style. The culture, values, and business styles should fit those of the financial institution. When a foreign based service provider is considered, the evaluation should assess the relationship in light of the above items as well as the information discussed in Appendix C, Foreign-Based Third-Party Service Providers.

Financial institutions may perform due diligence on one or more of the service providers that respond to the RFP. The depth and formality of the due diligence performed may vary according to the risk of the outsourced relationship, the institution's familiarity with the prospective service providers, and the stage of the provider selection process.

Once institutions issue RFPs, receive and evaluate responses, and perform due diligence, they enter into contract negotiations with one or more of the service providers they have determined can best meet their needs.

CONTRACT ISSUES

Action Summary

Before signing a contract, management should:

- Ensure the contract clearly defines the rights and responsibilities of both parties;
- Ensure the contract contains adequate and measurable service level agreements;
- Ensure contracts with affiliates clearly reflect an arms-length relationship and costs and services are at least as favorable to the institution as those available from a non-affiliated provider;
- Choose the most appropriate pricing method for the financial institution's needs;
- Ensure the contract does not contain provisions or inducements that may have a significant, adverse affect on the institution;
- Engage legal counsel to review the contract; and
- Evaluate foreign-based third-party service providers in light of the guidance found in this section and in Appendix C, Foreign-Based Third-Party Service Providers.

After selecting a service provider, management should negotiate a contract that meets their requirements. The RFP and the service provider's response can be used as inputs to this process. The contract is the legally binding document that defines all aspects of the servicing relationship. A written contract should be present in all servicing relationships. This includes instances where the service provider is affiliated with the institution. When contracting with an affiliate, the institution should ensure the costs and quality of services provided are commensurate with those of a nonaffiliated provider. The contract is the single most important control in the outsourcing process. Because of the importance of the contract, management should:

- Verify the accuracy of the description of the outsourcing relationship in the contract;
- Ensure the contract is clearly written and contains sufficient detail to define the rights and responsibilities of each party comprehensively; and
- Engage legal counsel early in the process to help prepare and review the proposed contract.

Examples of contract elements that should be considered include:

Scope of Service. The contract should clearly describe the rights and responsibilities of the parties to the contract. Considerations should include:

- Descriptions of required activities, timeframes for their implementation, and assignment of responsibilities. Implementation provisions should take into consideration other existing systems or interrelated systems to be developed by different service providers (e.g., an Internet banking system being integrated with existing core applications or systems customization);
- Obligations of, and services to be performed by, the service provider including software support and maintenance, training of employees, or customer service;
- Obligations of the financial institution;
- The contracting parties' rights in modifying existing services performed under the contract;[1] and
- Guidelines for adding new or different services and for contract re negotiation.

Performance Standards. Institutions should include performance standards that define minimum service level requirements and remedies for failure to meet standards in the contract. For example, common service level metrics include percent system uptime, deadlines for completing batch processing, or number of processing errors. Industry standards for service levels may provide a reference point. The institution should periodically review overall performance standards to ensure consistency with its goals and objectives. Also see the *Service Level Agreements* section in this booklet.

Security and Confidentiality. The contract should address the service provider's responsibility for security and confidentiality of the institution's resources (e.g., information, hardware).[2] The agreement should prohibit the service provider and its agents from using or disclosing the institution's information, except as necessary to or consistent with providing the contracted services, and to protect against

[1] Institutions may find advantages in contracting for services for three or more years because of the costs of entering into the contract, the costs of changing service providers, and favorable price breaks that may be offered by the vendor for longer terms. Contract flexibility is necessary under these circumstances because of the rapid changes occurring in an IT environment. Contract flexibility should allow for changes in service levels; increase or decrease in the scope of the process, service, or system due to changing institutional goals or objectives; and the retargeting of all relational elements on an annual basis. See Contract Inducement Concerns section in this booklet for further issues to be considered in entering into long-term contracts.

[2] The "Guidelines Establishing Standards to Safeguard Customer Information" to implement section 50I(b) of the Gramm–Leach–Bliley Act of I999 (GLBA) promulgated by the FFIEC agencies requires institutions to, among other things, require service providers by contract to implement appropriate security controls to comply with the guidelines with respect to their handling of customer information.

unauthorized use (e.g., disclosure of information to institution competitors). If the service provider receives nonpublic personal information regarding the institution's customers, the institution should verify that the service provider complies with all applicable requirements of the privacy regulations. Institutions should require the service provider to fully disclose breaches in security resulting in unauthorized intrusions into the service provider that may materially affect the institution or its customers. The service provider should report to the institution when intrusions occur, the effect on the institution, and corrective action to respond to the intrusion, based on agreements between both parties.

Controls. Management should consider implementing contract provisions that address the following controls:

- Service provider internal controls;
- Compliance with applicable regulatory requirements;
- Record maintenance requirements for the service provider;
- Access to the records by the institution;
- Notification requirements and approval rights for any material changes to services, systems, controls, key project personnel, and service locations;
- Setting and monitoring parameters for financial functions including payments processing or extensions of credit on behalf of the institution; and
- Insurance coverage maintained by the service provider.

Audit. The institution should include in the contract the types of audit reports It IS entitled to receive (e.g., financial, internal control, and security reviews). The contract should specify the audit frequency, any charges for obtaining the audits, as well as the rights of the institution and its regulatory agencies to obtain the results of the audits in a timely manner. The contract may also specify rights to obtain documentation of the resolution of any deficiencies and to inspect the processing facilities and operating practices of the service provider. Management should consider, based upon the risk assessment phase, if it can rely on internal audits or if there is a need for external audits and reviews.

For services involving access to open networks, such as Internet-related services, management should pay special attention to security. The institution should consider including contract terms requiring periodic control reviews performed by an independent party with sufficient expertise. These reviews may include penetration testing, intrusion detection, reviews of firewall configuration, and other independent control reviews. The institution should receive sufficiently detailed reports on the findings of these ongoing audits to assess security adequately without compromising the service provider's security.

Reports. Contractual terms should include the frequency and type of reports the institution will receive (e.g., performance reports, control audits, financial statements, security, and business resumption testing reports). The contracts should also outline the guidelines and fees for obtaining custom reports.

Business Resumption and Contingency Plans. The contract should address the service provider's responsibility for backup and record protection, including equipment, program and data files, and maintenance of disaster recovery and contingency plans. The contracts should outline the service provider's responsibility to test the plans regularly and provide the results to the institution. The institution should consider interdependencies among service providers when determining business resumption testing requirements. The service provider should provide the institution a copy of the contingency plan that outlines the required operating procedures in the event of business disruption. Contracts should include specific provisions for business recovery timeframes that meet the institution's business requirements. The institution should ensure that the contract does not contain any provisions that would excuse the service provider from implementing its contingency plans.

Sub-contracting and Multiple Service Provider Relationships. Some service providers may contract with third parties in providing services to the financial institution. Institutions should be aware of and approve all subcontractors. To provide accountability, the financial institution should designate the primary contracting service provider in the contract. The contract should also specify that the primary contracting service provider is responsible for the services outlined in the contract regardless of which entity actually conducts the operations. The institution should also consider including notification and approval requirements regarding changes to the service provider's significant subcontractors.

Cost. The contract should fully describe the calculation of fees for base services, including any development, conversion, and recurring services, as well as any charges based upon volume of activity or for special requests. Contracts should also address the responsibility and additional cost for purchasing and maintaining hardware and software. Any conditions under which the cost structure may be changed should be addressed in detail including limits on any cost increases. Also see the *Pricing Methods* and *Bundling* sections in this booklet.

Ownership and License. The contract should address the ownership, rights to, and allowable use of the institution's data, equipment/hardware, system documentation, system and application software, and other intellectual property rights. Ownership of the institution's data must rest clearly with the institution. Other intellectual property rights may include the institution's name and logo, its trademark or copyrighted material, domain names, web sites designs, and other work products developed by the service provider for the institution. Additional information regarding the development of customized software to support outsourced services can be found in the *IT Handbook's* "Development and Acquisition Booklet."

Duration. Institutions should consider the type of technology and current state of the industry when negotiating the appropriate length of the contract and its renewal periods. While there can be benefits to long-term technology contracts, certain technologies may be subject to rapid change and a shorter-term contract may prove beneficial. Similarly, institutions should consider the appropriate length of time required to notify the service provider of the institutions' intent not to renew the contract prior to expiration. Institutions should consider coordinating the expiration dates of contracts for inter-related services (e.g., web site, telecommunications, programming, network support) so that they coincide, where practical. Such coordination can minimize the risk of terminating a contract early and incurring penalties as a result of necessary termination of another related service contract.

Dispute Resolution. The institution should consider including a provision for a dispute resolution process that attempts to resolve problems in an expeditious manner as well as a provision for continuation of services during the dispute resolution period.

Indemnification. Indemnification provisions should require the service provider to hold the financial institution harmless from liability for the negligence of the service provider. Legal counsel should review these provisions to ensure the institution will not be held liable for claims arising as a result of the negligence of the service provider.

Limitation of Liability. Some service provider standard contracts may contain clauses limiting the amount of liability that can be incurred by the service provider. If the institution is considering such a contract, management should assess whether the damage limitation bears an adequate relationship to the amount of loss the financial institution might reasonably experience as a result of the service provider's failure to perform its obligations.

Termination. Management should assess the timeliness and expense of contract termination provisions. The extent and flexibility of termination rights can vary depending upon the service. Institutions should consider including termination rights for a variety of conditions including change in control (e.g., acquisitions and mergers), convenience, substantial increase in cost, repeated failure to meet service levels, failure to provide critical services, bankruptcy, company closure, and insolvency. The contract should establish notification and timeframe requirements and provide for the timely return of the institution's data and resources in a machine readable format upon termination. Any costs associated with conversion assistance should also be clearly stated.

Assignment. The institution should consider contract provisions that prohibit assignment of the contract to a third party without the institution's consent. Assignment provisions should also reflect notification requirements for any changes to material subcontractors.

Foreign-based service providers. Institutions entering into contracts with foreign-based service providers should consider a number of additional contract issues and provisions. See Appendix C included in this booklet.

Regulatory Compliance. Financial institutions should ensure that contracts with service providers include an agreement that the service provider and its services will comply with applicable regulatory guidance and requirements. The provision should also indicate that the service provider agrees to provide accurate information and timely access to the appropriate regulatory agencies based on the type and level of service it provides to the financial institution.

SERVICE LEVEL AGREEMENTS (SLAs)

Service level agreements are formal documents that outline the institution's predetermined requirements for the service and establish incentives to meet, or penalties for failure to meet, the requirements. Financial institutions should link SLAs to provisions in the contract regarding incentives, penalties, and contract cancellation in order to protect themselves against service provider performance failures.

Management should develop SLAs by first identifying the significant elements of the service. The elements can be related to tasks (i.e., processing error rates, system up-time, etc.) or they can be organizational (i.e., employee turnover). Once it has identified the elements, management should devise ways to measure the performance of those elements objectively. Finally, institutions should determine the frequency of the measurements and an acceptable range of results to determine when a service provider violates the SLA benchmarks.

Although the specific performance standards may vary with the nature of the service delivered, management should consider SLAs to address the following issues:

- Availability and timeliness of services;
- Confidentiality and integrity of data;
- Change control;
- Security standards compliance, including vulnerability and penetration management;
- Business continuity compliance; and
- Help desk support.

SLAs addressing business continuity should measure the service provider's or vendor's contractual responsibility for backup, record retention, data protection, and the maintenance of disaster recovery and contingency plans. The SLAs can also test the contingency plan's provisions for business recovery timeframes or conducting periodic tests of the plan. Neither contracts nor SLAs should contain any extraordinary provisions that would excuse the vendor or service provider from implementing its contingency plans (outsourcing contracts should include clauses that discuss unforeseen events for which the institution would not be able to adequately prepare).

PRICING METHODS

Financial institutions should have several choices when it comes to pricing an outsourcing venture. Management should consider all available pricing options and choose the most appropriate for the specific contract. Examples of different pricing methods include:

- Cost plus—The service provider receives payment for its actual costs, plus a predetermined profit margin or markup (usually percentage of actual costs).

For example, the service provider builds a website at a cost of $5,000 plus a 10% markup; the institution pays $5,500.

- Fixed price—The service provider price is the same for each billing cycle for the entire contract period. The advantage of this approach is that institutions know exactly what the provider will bill each month. Problems may arise if the institution does not adequately define the scope or the process. Often, with the fixed price method, the service provider labels services beyond the defined scope as additional or premium services. For example, if a service provider bills an institution $500 per month for maintaining a website, and the institution decides it wants to add another link, the service provider may charge more for that service if it is not clearly defined in the original contract.
- Unit pricing—The service provider sets a rate for a particular level of service, and the institution pays based on usage. For example, if an institution pays $.10 per hit on a website, and the site has 5,000 hits for the month, the institution pays $500 for the month.
- Variable pricing—The service provider establishes the price of the service based on a variable such as system availability. For example, the provider bills the institution $500, $600, or $800 per month for service levels of 99.00, 99.50, or 99.75 percent system availability, respectively. If a website was available 99.80 percent of the time in a billing period, the institution would pay $800.
- Incentive-based pricing—Incentives encourage the service provider to perform at peak level by offering a bonus if the provider performs well. This plan can also require the provider to pay a penalty for not performing at an acceptable level. For example, the institution wants a service provider to build a website. The service provider agrees to do so within 90 days for $5,000. The institution offers the provider $6,500 if the website is ready within 45 days, but states that it will only pay $3,500 if the provider fails to meet its 90 day deadline.
- Future price changes—Service providers typically include a provision that will increase costs in the future either by a specified percentage or per unit. Some institutions may also identify circumstances under which price reductions might be warranted (i.e., reduction in equipment costs).

BUNDLING

The provider may entice the institution to purchase more than one system, process, or service for a single price – referred to as "bundling." This practice may result in the institution getting a single consolidated bill that may not provide information relating to pricing for each specific system, process, or service. Although the bundled services may appear to be cheaper, the institution cannot analyze the costs of the individual services. Bundles may include processes and services that the institution does not want or need. It also may not allow the institution to discontinue a specific system, process, or service without having to renegotiate the contract for all remaining services.

CONTRACT INDUCEMENT CONCERNS

Financial institutions should not sign servicing contracts that contain provisions or inducements that may adversely affect the institution. Such contract provisions may include extended terms (up to 10 years), significant increases in costs after the first few years, and/or substantial cancellation penalties. In addition, some service contracts improperly offer inducements that allow an institution to retain or increase capital by deferring losses on the disposition of assets or avoiding expense recognition. These inducements may attract institutions wanting to mask capital problems.

Inducements can take several forms including the following examples:

- The service provider purchases certain assets (e.g., computer equipment or foreclosed real estate) at book value (which exceeds market value) or purchases capital stock from the institution.
- The service provider offers cash bonuses to the institution upon completion of the conversion.
- The service provider offers up-front cash to the institution. The provider states that the institution acquires the right to future cost savings or profit enhancements that will accrue to the institution because of greater operational efficiencies. These improvements are usually without measurable benchmarks.
- The institution defers expenses for conversion costs or processing fees under the terms of the contract.
- Low installation and conversion costs in exchange for higher future systems support and maintenance costs.

These inducements may offer a short-term benefit to the institution. However, the provider usually recoups the costs by charging a premium for the processing services. These excessive fees may adversely affect an institution's financial condition over the long-term. Furthermore, institutions should account for such inducements in accordance with generally accepted accounting principles (GAAP) and regulatory reporting requirements.

Accordingly, when negotiating contracts, an institution should ensure the provider furnishes a level of service that meets the needs of the institution over the life of the contract. The institution must ensure it accounts for contracts in accordance with GAAP. Contracting for excessive servicing fees and/or failing to account properly for such transactions is an unsafe and unsound practice. In entering into service agreements, institutions must ensure accounting under such agreements reflects the substance of the transaction and not merely the form.

ONGOING MONITORING

Action Summary

Management should monitor service provider performance and potential changes in institution requirements throughout the life of the contract. Monitoring should encompass:

- Key service level agreements (SLAs) and contract provisions;
- Financial condition of the service provider;
- General control environment of the service provider through the receipt and review of audit reports and other internal control reviews; and
- Potential changes due to the external environment.

Financial institutions should have an oversight program to ensure service providers deliver the quantity and quality of services required by the contract. The monitoring program should target the key aspects of the contracting relationship with effective monitoring techniques. The program should monitor the service provider environment including its security controls, financial strength, and the impact of any external events. The resources to support this program will vary depending on the criticality and complexity of the system, process, or service being outsourced.

To increase monitoring effectiveness, management should periodically rank service provider relationships according to risk to determine which service providers require closer monitoring. Management should base the rankings on the residual risk of the relationship after analyzing the quantity of risk relative to the controls over those risks. Relationships with higher risk ratings should receive more frequent and stringent monitoring for due diligence, performance (financial and/or operational), and independent control validation reviews. Personnel responsible for provider oversight should have the necessary expertise to assess the risks and should maintain suitable documentation. Management should use the oversight documentation when renegotiating contracts as well as developing contingency planning requirements.

User groups are another mechanism financial institutions can use to monitor and influence their service provider. User groups can participate and influence service provider testing (i.e., security, disaster recovery, and systems) as well as promote client issues. Independent user groups can monitor and influence a service provider better than its individual clients. Collectively, the group will constitute a significant portion of the service provider's business.

KEY SERVICE LEVEL AGREEMENTS AND CONTRACT PROVISIONS

Management should include SLAs in its outsourcing contracts to specify and clarify performance expectations, as well as establish accountability. These SLAs formalize the performance criteria against which the quantity and quality of service should be measured. Management should closely monitor the service provider's compliance with key service level agreements. To ensure an effective oversight program, the institution should develop:

- A formal policy that defines the SLA program;
- An SLA monitoring process;
- A recourse process for non-performance;
- An escalation process;
- A dispute resolution process; and
- A termination process.

FINANCIAL CONDITION OF SERVICE PROVIDERS

Institutions should have on-going monitoring of the financial condition of their provider(s). To fulfill its fiduciary responsibility, an institution involved in an outsourcing arrangement should determine the financial viability of its provider(s) on an annual basis. However, if the financial condition of the provider is declining or unstable, more frequent financial reviews are warranted. Once the financial review is complete, management should report the results to the board of directors or to a designated committee. At a minimum, management's review should contain a careful analysis of the provider's annual financial statement. Institution management may also use other forms of information to determine a provider's condition, such as independent auditor reports. These reports may contain information that can be vital in determining a provider's financial condition. Managers also can use information provided by public media (trade magazines, newspapers, television, etc.).

If the institution becomes aware that the provider's financial condition is unstable or deteriorating, the institution should implement its contingency plan. Even if the provider remains in operation, its financial problems may jeopardize the quality of its service and possibly the integrity of the data in its possession. Institutions should consider a provider's failure to provide adequate financial data as a potential red flag that there may be serious financial stability issues.

Termination of services due to the bankruptcy of the service provider can have a devastating effect on a serviced institution's operations. There may not be sufficient advance notice of termination, an effective contingency plan, or adequate access to provider personnel. In such a situation, the serviced institution is put into the position of having to find an alternate processing site with little advance notice.

At this point, a serviced institution has several alternatives including:

- Paying off the servicer's creditor(s) and hiring outside specialists to operate the center;
- Obtaining required equipment and software for in-house processing; and
- Transferring data files to another provider.

Most options are costly and may cause harmful operating delays.

In some instances, the provider owns the programs and documentation required to process the institution's files. Unless the contract contains an escrow agreement for source code, the program and documentation are unavailable to the institution. These programs are often the TSPs only significant assets. Therefore, a creditor of a bankrupt TSP, in an attempt to recover outstanding debts, might seek to attach those assets and further limit their availability to institutions. The bankruptcy court may provide remedies to the institution, but only after adjudicating substantive matters.

GENERAL CONTROL ENVIRONMENT OF THE SERVICE PROVIDER

To oversee the risks associated with the use of external providers effectively, the institution should evaluate the adequacy of a provider's internal and security controls. Management should ensure the provider develops and adheres to appropriate policies, procedures, and standards. When conducting its evaluation, the institution should consider the results of internal audits conducted by institution staff or a user group, as well as external audits and control reviews conducted by qualified sources The *IT Handbook's* "Audit Booklet" provides additional details on the various types of external audit engagements for third-party audits of a service provider.

The institution's review of the audit should include an assessment of the following factors in order to determine the adequacy of a service provider's internal and security controls:

- The practicality of the service provider having an internal auditor, and the auditor's level of training and experience;
- The service providers external auditors' training and background; and
- Internal IT audit techniques of the service provider.

Financial institutions should conduct a regular, comprehensive audit of their service provider relationships. The audit scope should include a review of controls and operating procedures that help protect the institution from losses due to irregularities and willful manipulations.

SAS 70 reports generated on external providers typically identify certain internal control measures that client institutions are responsible for implementing in order for the provider's accounting systems to be effective. These client institution internal control measures are essential. Financial institution management and audit personnel should verify that the recommended institution internal controls are working effectively, and that the controls effectively complement the accounting system controls described in the provider's third-party review.

Because of the need for an effective internal control program, designated personnel should periodically perform "around-the-computer" audit techniques that:

- Develop data controls (proof totals, batch totals, document counts, number of accounts, and pre-numbered documents) at the institution before submission to the provider. The auditor should sample the controls periodically to ensure their accuracy.
- Include spot-checking reconcilement procedures to ensure output totals agree with input totals, less any rejects.
- Sample rejected, un-posted, holdover, and suspense items to determine why they did not process and how they are addressed (to assure they are properly corrected and reentered on a timely basis).
- Verify selected master file information (such as service charge codes), review exception reports, and crosscheck loan extensions and deposit account entries to source documents.
- Spot-check computer calculations, such as loan rebates, interest on deposits, late charges, service charges, and past-due loans.
- Trace transactions to final disposition to ensure there are adequate audit trails.
- Review source input to ensure sensitive master-file change requests have the required prior approval by appropriate staff or management.
- Visit the provider periodically to assess the status of controls.
- Review other provider audits.

In addition, "through-the-computer" audit techniques allow the auditor to use the computer to check processing steps. These techniques use audit software programs to test extensions and footings and to prepare direct verification statements. These audit software programs often can invoke statistical sampling routines in generating their audit confirmations. If a serviced institution has audit software, it should make arrangements with the provider to allow its use.

Regardless of whether the information processing is internal or outsourced, the financial institution's board of directors should ensure adequate audit coverage. If the institution has no technical audit expertise, the non-technical audit methods can provide minimum coverage. The institution should supplement the internal audit with comprehensive outside IT audits.

POTENTIAL CHANGES DUE TO THE EXTERNAL ENVIRONMENT

The contract between the institution and the service provider should be written to encompass the institution's requirements at the time the contract is formed. Over time, the institution's needs may change due to changes in regulation, the economic environment, competition, and other factors outside the contract. Although the contract should provide for flexibility to meet those changing needs, the institution should monitor for changes and update its contract accordingly.

RELATED TOPICS

Action Summary

Financial institutions should:

- Establish ongoing and effective business continuity and information security monitoring programs;
- Effectively manage multiple service provider relationships; and
- Assess, monitor, and effectively control cross-border risks when foreign service providers are used.

BUSINESS CONTINUITY PLANNING

Each financial institution should have an effective business continuity plan as outlined in the *IT Handbook's* "Business Continuity Planning Booklet." The financial institution should also establish ongoing effective business continuity monitoring programs to ensure TSPs adequately control the risks, including information security aspects, associated with the technology services provided. The financial institution has responsibility not only for those portions of the business continuity program performed in-house, but for any portions of the plan developed by a service provider or otherwise outsourced. Financial institutions should consider TSP-related business continuity programs when developing internal plans and programs.

The outsourcing risk management program should identify, for Business Continuity Planning (BCP) purposes, the specific responsibilities of all parties, particularly in the areas of information security and business continuity planning. Financial institutions must also consider which of their critical financial services rely on TSP services, including key telecommunication and network service providers.

The institution should understand all relevant service provider business continuity requirements, incorporate those requirements within its own business continuity plan, and ensure the service provider tests its plan annually. Management should require the service provider to report all test plan results and to notify the institution after any business continuity plan modifications. The institution should integrate the provider's business continuity plan into its own plan, communicate functions to the appropriate personnel, and maintain and periodically review the combined plan.

Many financial institutions rely on outside data processing providers and any extended interruption or termination of service can disrupt normal operations. Termination of services should occur according to the terms of the service contract, but can result from unanticipated events.

If the provider complies with basic industry standards and maintains an effective business continuity plan, disruption of services should be minimal and the contract will remain intact. The business continuity plan should require the provider to maintain current data files and programs at an alternative site and arrange for processing at another location. At a minimum, these provisions should allow the provider to process the most important data applications. The institution's business continuity plan, which should complement the provider's plan, is an essential recovery tool when disruption occurs with minimal advance notice.

Events that can cause interruption in the availability of an institution's technology include natural disasters, accidents, software errors, hardware failure, utility outages, and social, political, and economic instability. Even with an outsourcing arrangement, the institution should ensure appropriate backup provisions have been established for their critical data and related processing functions. Effective backup procedures will allow the institution to continue processing applications in the event the data communication system fails. Numerous options are available for management to consider, such as using batch rather than real-time processing methods, operating PCs in an offline mode, capturing data at the controller if transmission lines are lost, or altering communication links through redundant data communication lines, backup modems, or rerouted circuits from the local telephone carrier. Institutions that perform data capture or other functions in-house, should address alternative sites or other means in their backup plan to recover or continue these functions.

Regardless of the method used, an institution should have a comprehensive backup plan with procedures that detail how to obtain and use personnel and equipment. Institutions should test backup capabilities periodically to ensure protection is available and employees are familiar with the plan.

With respect to monitoring and maintaining business continuity plans, institutions should:

- Regularly review the business continuity plans of the service provider or vendor to ensure any services considered "mission critical" for the financial institution could be restored within an acceptable timeframe.
- Review the service provider's program for contingency plan testing. For critical services, annual or more frequent tests of the contingency plan are required.
- Assess service provider/vendor interdependencies for mission critical services and applications.

OUTSOURCING THE BUSINESS CONTINUITY FUNCTION

In addition to ensuring that outsourced financial and technology services include appropriate business continuity plans; financial institutions that outsource all or a portion of their business continuity capability should consider the following factors.

- *Staffing*—The provider should have sufficient and knowledgeable staff available to provide appropriate onsite technical support to ensure timely resumption of operations at the recovery site.
- *Processing Time Availability*—The provider should allocate sufficient processing time, resources, and security controls to accommodate the potential for multiple clients. The institution should ensure it could process normal volumes of work within appropriate time requirements.
- *Access Rights*—The provider should disclose any access limitations. The provider should guarantee the institution's right to use the site in case of an emergency. Alternatively, the institution should understand any priority arrangements. For example, some sites operate on a first-come, first-serve basis until the site is at full capacity, but others have pre-arranged priorities based on contractual agreements.
- *Hardware and Software*—The recovery site should have compatible hardware and software. The institution should monitor the compatibility of the site to handle its specific computer hardware and software requirements. To facilitate the monitoring, the provider should be required by contract to notify the institution of any changes in the hardware, software, and equipment at the recovery site.
- *Security Controls*—The institution should ensure it can maintain adequate physical and logical security controls at the recovery site.
- *Testing*—The service provider contract should address access to the recovery site for periodic testing. At a minimum, the institution needs sufficient access to perform at least one full-scale test of the recovery site annually, including verification of telecommunications capabilities. Similarly, the institution should ensure the service provider also performs periodic tests of its own BCP and submits test results to customer financial institutions.
- *Confidentiality of Data*—The institution should ensure the provider can maintain the confidentiality of its business and customer data. The service provider should maintain controls sufficient to ensure the security and confidentiality of the information assets consistent with the institution's information security program. Confidentiality of data is particularly important when multiple clients operate from the same recovery site. Institution management should establish whether the service provider has addressed these issues in its contract, particularly the provisions concerning the Interagency Guidelines Establishing Standards for Safeguarding Customer Information.[1]
- *Telecommunications*—The institution should review telecommunications redundancy and capacity at the recovery site, including how communications from the institutions to the recovery site will be established. The service

[1] See 66 Federal Register 8616 (Feb. 1, 2001); 12 CFR Part 30, app. B (OCC); 12 CFR Part 208, app. D-2 and Part 225, app. F. (Board); 12 CFR Part 364, app. B (FDIC); 12 CFR Part 570, app. B (OTS). See 66 Federal Register 8152 (Jan. 30, 2001); 12 CFR Part 748, app. A (NCUA).

provider should take steps to ensure the recovery site will have adequate telecommunications services (both voice and data) for all of its clients.
- *Reciprocal Agreements*—Financial institutions contracting with another institution for a recovery site should consider the above issues of staffing, processing availability, access rights for recovery or testing, compatibility, security, capacity, etc. Both institutions should ensure they maintain sufficient capacity to meet recovery time objectives and minimum service levels in the event one institution needs to recover operations
- *Space*—The recovery site should have adequate space to accommodate the affected institution's recovery staff.
- *Printing Capacity/Capability*—The recovery site should maintain adequate printing capacity to meet the demand of the affected institution under acceptable levels of service.
- *Contacts*—Institution management should know the procedures for declaring a disaster including who has the authority to declare a disaster and initiate use of the recovery site. Also, the institution should maintain an updated list of contacts names and numbers for the recovery site provider and know the procedures for communicating with the provider.

Outsourced business continuity arrangements can be cost-effective for smaller institutions when compared to establishing and maintaining dedicated alternate recovery sites. Institutions should periodically conduct a thorough test of outsourced disaster recovery services (at least annually).

INFORMATION SECURITY/SAFEGUARDING

Information assets are valuable, and institutions should ensure these assets are adequately protected in outsourcing relationships. Financial institutions have a legal responsibility to ensure service providers take appropriate measures designed to meet the objectives of the information security guidelines, and comply with GLBA 501 (b). Those measures should result from the institution's security process and should be included or referenced in the contract between the institution and the service provider. Refer to the IT Handbook's "Information Security Booklet" for additional information on the information security process.

In choosing service providers, management should exercise appropriate due diligence to ensure the protection of both financial institution and customer assets. Before entering into outsourcing contracts, and throughout the life of the relationship, institutions should ensure the service provider's physical and data security standards meet or exceed standards required by the institution. Institutions should also implement adequate protections to ensure service providers and vendors are only given access to the information and systems that they need to perform their function. Management should restrict their access to financial institution systems, and appropriate access controls and monitoring should be in place between service provider's systems and the institution.

MULTIPLE SERVICE PROVIDER RELATIONSHIPS

A multiple service provider relationship is an environment where two or more service providers collaborate to deliver an end-to-end solution to the financial institution.

An institution can select from two techniques to manage this relationship, but remains responsible for understanding and monitoring the control environment of all servicers that have access to the financial institution's systems, records, or resources. The first technique involves the use of a lead service provider to manage the institution's various technology providers. The second technique, which may present its own set of implementation challenges, involves the use of operational agreements between each of the service providers or stand-alone contracts. If the first technique is employed, management should ensure its primary service provider has a contractual obligation to notify the financial institution of any concerns (controls/performance) associated with any of its outsourced activities. Management should also ensure the service provider's control environment meets or exceeds the institution's expectations, including the control environment of organizations that the primary service provider utilizes.

Stand-alone contracts with each service provider require increased management of each provider. Contracting for a technology solution by using one lead provider may lessen the need for the institution to become directly involved if subcontractors fail to perform, but it does not diminish the responsibility for monitoring the internal and security controls of subcontractors through the primary service provider relationship. Because the institution has less control using the lead provider approach, management should require by contract that TSPs notify the institution of all subcontractor relationships.

OUTSOURCING TO FOREIGN SERVICE PROVIDERS

Some institutions develop outsourcing relationships with service providers located in foreign countries. These arrangements can provide cost, expertise, and other advantages to the institutions and should be subject to the same due diligence and assessment as domestic outsourcing relationships. In addition, foreign outsourcing relationships result in unique strategic, reputation, credit, liquidity, transactional, geographic, and compliance risks that institutions should identify, assess, prevent, and control. See Appendix C for additional detail.

APPENDIX A: EXAM PROCEDURES

EXAMINATION OBJECTIVE: Assess the effectiveness of the institution's risk management process as it relates to the outsourcing of information systems and technology services.

- Tier I objectives and procedures relate to the institution's implementation of a process for identifying and managing outsourcing risks.
- Tier II objectives and procedures provide additional validation and testing techniques as warranted by risk to verify the effectiveness of the institution's process on individual contracts.

Tier I and Tier II are intended to be a tool set examiners will use when selecting examination procedures for their particular examination. Examiners should use these procedures as necessary to support examination objectives.

TIER I OBJECTIVES AND PROCEDURES

Objective I: Determine the appropriate scope for the examination.

1. Review past reports for weaknesses involving outsourcing. Consider:
   - Regulatory reports of examination of the institution and service provider(s); and
   - Internal and external audit reports of the institution and service provider(s) (if available).
2. Assess management's response to issues raised since the last examination. Consider:
   - Resolution of root causes rather than just specific issues; and
   - Existence of any outstanding issues.
3. Interview management and review institution information to identify:
   - Current outsourcing relationships and changes to those relationships since the last examination. Also identify any:
     - Material service provider subcontractors,
     - Affiliated service providers,
     - Foreign-based third party providers;
   - Current transaction volume in each function outsourced;
   - Any material problems experienced with the service provided;
   - Service providers with significant financial or control related weaknesses; and
   - When applicable, whether the primary regulator has been notified of the outsourcing relationship as required by the Bank Service Company Act or Home Owners' Loan Act.

Objective 2: Evaluate the quantity of risk present from the institution's outsourcing arrangements.

1. Assess the level of risk present in outsourcing arrangements. Consider risks pertaining to:
   - Functions outsourced;
   - Service providers, including, where appropriate, unique risks inherent in foreign-based service provider arrangements; and
   - Technology used.

Objective 3: Evaluate the quality of risk management

1. Evaluate the outsourcing process for appropriateness given the size and complexity of the institution. The following elements are particularly important:
   - Institution's evaluation of service providers consistent with scope and criticality of outsourced services; and
   - Requirements for ongoing monitoring.

2. Evaluate the requirements definition process.
   - Ascertain that all stakeholders are involved; the requirements are developed to allow for subsequent use in request for proposals (RFPs), contracts, and monitoring; and actions are required to be documented; and
   - Ascertain that the requirements definition is sufficiently complete to support the future control efforts of service provider selection, contract preparation, and monitoring.
3. Evaluate the service provider selection process.
   - Determine that the RFP adequately encapsulates the institution's requirements and that elements included in the requirements definition are complete and sufficiently detailed to support subsequent RFP development, contract formulation, and monitoring;
   - Determine that any differences between the RFP and the submission of the selected service provider are appropriately evaluated, and that the institution takes appropriate actions to mitigate risks arising from requirements not being met; and
   - Determine whether due diligence requirements encompass all material aspects of the service provider relationship, such as the provider's financial condition, reputation (e.g., reference checks), controls, key personnel, disaster recovery plans and tests, insurance, communications capabilities and use of subcontractors.
4. Evaluate the process for entering into a contract with a service provider. Consider whether:
   - The contract contains adequate and measurable service level agreements;
   - Allowed pricing methods do not adversely affect the institution's safety and soundness, including the reasonableness of future price changes;

- The rights and responsibilities of both parties are sufficiently detailed;
- Required contract clauses address significant issues, such as financial and control reporting, right to audit, ownership of data and programs, confidentiality, subcontractors, continuity of service, etc.;
- Legal counsel reviewed the contract and legal issues were satisfactorily resolved; and
- Contract inducement concerns are adequately addressed.

5. Evaluate the institution's process for monitoring the risk presented by the service provider relationship. Ascertain that monitoring addresses:
   - Key service level agreements and contract provisions;
   - Financial condition of the service provider;
   - General control environment of the service provider through the receipt and review of appropriate audit and regulatory reports;
   - Service provider's disaster recovery program and testing;
   - Information security;
   - Insurance coverage;
   - Subcontractor relationships including any changes or control concerns;
   - Foreign third party relationships; and
   - Potential changes due to the external environment (i.e., competition and industry trends).
6. Review the policies regarding periodic ranking of service providers by risk for decisions regarding the intensity of monitoring (i.e., risk assessment). Decision process should:
   - Include objective criteria;
   - Support consistent application;
   - Consider the degree of service provider support for the institution's strategic and critical business needs, and
   - Specify subsequent actions when rankings change.
7. Evaluate the financial institution's use of user groups and other mechanisms to monitor and influence the service provider.

Objective 4: Discuss corrective action and communicate findings

1. Determine the need to complete Tier II procedures for additional validation to support conclusions related to any of the Tier I objectives.
2. Review preliminary conclusions with the EIC regarding:
   - Violations of law, rulings, regulations;
   - Significant issues warranting inclusion in the Report as matters requiring attention or recommendations; and
   - Potential impact of your conclusions on the institution's risk profile and composite or component IT ratings.
3. Discuss findings with management and obtain proposed corrective action for significant deficiencies.

4. Document conclusions in a memo to the EIC that provides report ready comments for the Report of Examination and guidance to future examiners.
5. Organize work papers to ensure clear support for significant findings by examination objective.

TIER II OBJECTIVES AND PROCEDURES

A. IT REQUIREMENTS DEFINITION

1. Review documentation supporting the requirements definition process to ascertain that it appropriately addresses:
   - Scope and nature;
   - Standards for controls;
   - Minimum acceptable service provider characteristics;
   - Monitoring and reporting;
   - Transition requirements;
   - Contract duration, termination, and assignment' and
   - Contractual protections against liability.

B. DUE DILIGENCE

1. Assess the extent to which the institution reviews the financial stability of the service provider:
   - Analyzes the service provider's audited financial statements and annual reports;
   - Assesses the provider's length of operation and market share;
   - Considers the size of the institution's contract in relation to the size of the company;
   - Reviews the service provider's level of technological expenditures to ensure on going support; and
   - Assesses the impact of economic, political, or environmental risk on the service provider's financial stability.
2. Evaluate whether the institution's due diligence considers the following:
   - References from current users or user groups about a particular vendor's reputation and performance;
   - The service provider's experience and ability in the industry;
   - The service provider's experience and ability in dealing with situations similar to the institution's environment and operations;
   - The cost for additional system and data conversions or interfaces presented by the various vendors;
   - Shortcomings in the service provider's expertise that the institution would need to supplement in order to fully mitigate risks;

- The service provider's proposed use of third parties, subcontractors, or partners to support the outsourced activities;
- The service provider's ability to respond to service disruptions;
- Key service provider personnel that would be assigned to support the institution;
- The service provider's ability to comply with appropriate federal and state laws. In particular, ensure management has assessed the providers' ability to comply with federal laws (including GLBA and the USA PATRIOT Act[1]); and
- Country, state, or locale risk.

C. SERVICE CONTRACT

1. Verify that legal counsel reviewed the contract prior to closing.
   - Ensure that the legal counsel is qualified to review the contract particularly if it is based on the laws of a foreign country or other state; and
   - Ensure that the legal review includes an assessment of the enforceability of local contract provisions and laws in foreign or out-of-state jurisdictions.
2. Verify that the contract appropriately addresses:
   - Scope of services;
   - Performance standards;
   - Pricing;
   - Controls;
   - Financial and control reporting;
   - Right to audit;
   - Ownership of data and programs;
   - Confidentiality and security;
   - Regulatory compliance;
   - Indemnification;
   - Limitation of liability;
   - Dispute resolution;
   - Contract duration;
   - Restrictions on, or prior approval for, subcontractors;
   - Termination and assignment, including timely return of data in a machinereadable format;
   - Insurance coverage;
   - Prevailing jurisdiction (where applicable);
   - Choice of Law (foreign outsourcing arrangements);
   - Regulatory access to data and information necessary for supervision; and
   - Business Continuity Planning.

[1] Pub. L. No. 107-56 (Oct. 26, 2001).

3. Review service level agreements to ensure they are adequate and measurable. Consider whether:
   - Significant elements of the service are identified and based on the institution's requirements;
   - Objective measurements for each significant element are defined;
   - Reporting of measurements is required;
   - Measurements specify what constitutes inadequate performance; and
   - Inadequate performance is met with appropriate sanctions, such as reduction in contract fees or contract termination.
4. Review the institution's process for verifying billing accuracy and monitoring any contract savings through bundling.

D. MONITORING SERVICE PROVIDER RELATIONSHIP(S)

1. Evaluate the institution's periodic monitoring of the service provider relationship(s), including:
   - Timeliness of review, given the risk from the relationship;
   - Changes in the risk due to the function outsourced;
   - Changing circumstances at the service provider, including financial and control environment changes;
   - Conformance with the contract, including the service level agreement; and
   - Audit reports and other required reporting addressing business continuity, security, and other facets of the outsourcing relationship.
2. Review risk rankings of service providers to ascertain
   - Objectivity;
   - Consistency; and
   - Compliance with policy.
3. Review actions taken by management when rankings change, to ensure policy conformance when rankings reflect increased risk.
4. Review any material subcontractor relationships identified by the service provider or in the outsourcing contracts. Ensure:
   - Management has reviewed the control environment of all relevant subcontractors for compliance with the institution's requirements definitions and security guidelines; and
   - The institution monitors and documents relevant service provider subcontracting relationships including any changes in the relationships or control concerns.

APPENDIX B: REFERENCES

LAWS

- 12 USC 1464 (d) (7) Home Owners' Loan Act (Thrifts)
- 12 USC 1867 (c) (11), Bank Service Company Act (Banks)
- 15 USC 6801, Gramm-Leach-Bliley Act
- Pub. L. No. 107-56, USA PATRIOT Act

FEDERAL DEPOSIT INSURANCE CORPORATION

GUIDANCE

- FIL-49-99: Bank Service Company Act (June 3, 1999)
- FIL-50-2001: Bank Technology Bulletin: Technology Outsourcing Information Documents (June 4, 2001)

Attachments:

"Effective Practices for Selecting a Service Provider"
"Tools to Manage Technology Providers' Performance Risk: Service Level Agreements"
"Techniques for Managing Multiple Service Providers"

FEDERAL RESERVE BOARD

GUIDANCE

- SR 00-4 (SUP), Outsourcing of Information and Transaction Processing (February 29, 2000)

NATIONAL CREDIT UNION ADMINISTRATION

GUIDANCE

- NCUA Letter to Credit Unions No. 02-CU-17: E-Commerce Guide for Credit Unions (December 2002)
- NCUA Letter to Credit Unions No. 01-CU-20: Due Diligence Over Third Party Service Providers (November 2001)

OFFICE OF THE COMPTROLLER OF THE CURRENCY

GUIDANCE

- OCC Bulletin 2002-16: Bank Use of Foreign-Based Third-Party Service Providers (May 15, 2002)

- OCC Bulletin 2001-47, Third-Party Relationships, Risk Management Principles

OFFICE OF THRIFT SUPERVISION

Regulations

- 12 CFR Part 570, Appendix B: Interagency Guidelines Establishing Standards for Safeguarding Customer Information

GUIDANCE

- Thrift Bulletin 82: Third Party Arrangements (March 18, 2003)
- CEO Letter 113: Internal Controls (July 14, 1999)
- Thrift Activities Handbook: Section 340, Internal Control
- Thrift Activities Handbook: Section 341, Technology Risk Controls

APPENDIX C: FOREIGN-BASED THIRD PARTY SERVICE PROVIDERS

The material provided in this appendix focuses on foreign-based third-party service providers and should be used, in addition to all other material in this booklet, when examining such relationships. This appendix discusses the primary risks that may arise from service relationships between financial institutions and foreign-based third-parties[1], the steps institutions should consider when managing those risks, and the implications of the relationships within the context of the examination process.

BACKGROUND

Organizations often use domestic third-party service providers as an economic alternative to internal technology and data processing functions. Increasingly, these organizations are considering arrangements with foreign-based third parties or domestic firms that subcontract portions of their operations to foreign-based entities.

The use of foreign-based service providers is a common business practice that can be a less costly alternative to self-processing or to using domestic service providers. However, this practice raises country, compliance, contractual, reputation, operational (e.g., transactional), and strategic issues in addition to those presented by use of a domestic service provider. In managing these issues, management should conduct appropriate risk assessments and due diligence procedures and closely evaluate all contracts. Additionally, management should establish ongoing monitoring and oversight procedures.

RISK MANAGEMENT

A financial institution's senior managers are responsible for understanding the risks associated with foreign-based relationships and for ensuring that effective risk management practices are in place. Management should determine if a foreign-based technology relationship is consistent with the organization's overall business and technology strategies and if it can mitigate identified risks adequately. Before management executes a contract with foreign-based entities, it should consider issues

[1] The terms "foreign-based third-party service providers" or "foreign-based service provider" refer to any entity, including an affiliated organization or holding company, whose servicing operations are located in and subject to the laws of any country other than the United States, including service providers located outside the United States providing services to foreign branches of U .S. organizations. The term also includes the foreign operations, whether by subcontract or otherwise, of a domestic service provider.

such as choice-of-law and jurisdictional considerations. Additionally, organizations should establish appropriate due diligence and risk management policies that include oversight and monitoring procedures. These policies and procedures should consider that all of the risks associated with domestic third party providers are present in foreign-based arrangements in addition to the unique issues such as country and compliance risks arising from the fact that the third parties may not fall under the jurisdiction of domestic laws and regulations.

COUNTRY RISK

Country risk is an exposure to economic, social, and political conditions in a foreign country that could adversely affect a vendor's ability to meet its service level requirements. In certain situations, country risks could result in the loss of an organization's data, research, or development efforts. Managing country risk requires organizations to gather and assess information regarding foreign political, social, and economic conditions and events, and to address the exposures introduced by the relationship with a foreign-based provider. Risk management procedures should include the establishment of contingency, service continuity, and exit strategies in the event of unexpected disruptions in service.

COMPLIANCE RISK

Compliance risk involves the impact foreign-based arrangements could have on an organization's compliance with applicable U.S. and foreign laws and regulations. An organization's use of a foreign-based third party service provider should not inhibit the organization's compliance with applicable U.S. laws including consumer protection, privacy (Section 501(b) of GLBA/)[1], and information security laws as well as Bank Secrecy Act requirements[2] concerning the reporting and documentation of financial transactions. Additionally, organizations should consider the impact and operational requirements of foreign data privacy laws or regulatory requirements.[3] Organizations engaging foreign-based entities should also consider the sanctions

---

[1] 15 USC 6801. Gramm-Leach-Biiley Act, Section 501(b).

[2] In this regard, organizations using foreign-based service providers should be aware of Section 319 of the USA Patriot Act, Pub. L. No. 107-56 (Oct. 26, 2001), which requires a financial institution to make information on anti-money laundering compliance by the institution or its customers available within 120 hours of a government request.

[3] Organizations should identify and understand the application of any laws within a foreign jurisdiction that apply to information transferred from the United States to that foreign jurisdiction over the Internet or otherwise to information transferred from that jurisdiction to the United States, as well as to information collected within the foreign jurisdiction using automated or other equipment in that jurisdiction.

and embargo provisions[1] of the U.S. Treasury Office of Foreign Assets Control (OFAC) as well as the requirements regarding exportation of encryption-related technologies discussed in the following paragraph.

Export Controls

The United States has export control laws that restrict the export of software and other items (U.S. Export Administration Regulations).[2] These laws apply to all aspects of encryption usage, including but not limited to, software, hardware, and network applications. Organizations should ensure they and their service provider(s) comply with these laws. Contracts should include a representation and warranty that service providers will comply with U.S. export control laws.

DUE DILIGENCE

Management of an organization considering a foreign-based outsourcing arrangement should perform appropriate due diligence similar to domestic outsourcing arrangements before selecting or contracting with a service provider. The process should include an evaluation of a firm's financial stability and commitment to service, and the potential impact of the foreign jurisdiction's regulations, laws, accounting standards, and business practices. Additionally, management should consider the degree to which geographic distance, language, or social, economic, or political changes may affect the foreign-based service provider's ability to meet the organization's servicing needs. Management should consider the cost and logistical implications of managing a cross-border relationship, including the ongoing costs of managing and monitoring cross-border and foreign-based provider relationships.

CONTRACTS

Contracts between an organization and a foreign-based entity should address the risks identified during risk assessments and due diligence processes. Specific topics that should be considered regarding such contracts are discussed in the following paragraphs.

---

1 The Office of Foreign Assets Control of the U.S. Department of the Treasury administers and enforces economic and trade sanctions against certain foreign countries, organizations sponsoring terrorism, and international narcotics traffickers based on U.S. foreign policy and national security goals. For more information, refer to the OFAC Web site at www.treas.gov/ofac.

2 Export controls on commercial encryption products are administered by the Bureau of Industry and Security, part of the Department of Commerce. Organizations may be exporters if they provide encryption software to a foreign-based service provider, but some exceptions are available that apply to foreign national employees, including contractors and consultants, of U.S. companies and their subsidiaries inside and outside the United States. Export administration regulations regarding encryption are contained in 15 CFR §§ 740.13, 740.17 & 742.15. See www.bis.doc.gov.

Security, Confidentiality and Ownership of Data
Management should require contract provisions to protect its customers' privacy and the confidentiality of organizational records in conformance with U.S. laws and regulations. Federal regulations require that service provider contracts include provisions requiring the service provider to implement procedures and security measures that meet the objectives of customer information security guidelines.[1] Additionally, contracts should include provisions prohibiting the disclosure of any customer information to nonaffiliated third parties, other than as permitted under U.S. privacy laws.[2]

Any agreement with a foreign-based service provider should also include a provision that all information transferred to the foreign-based entity remains the property of the organization, regardless of how it is processed, stored, copied, or reproduced.

Regulatory Authority
Arrangements with foreign-based service providers should contain a provision acknowledging the authority of U.S. regulatory authorities[3] (pursuant to the Bank Service Company Act or the Home Owner's Loan Act) to examine the services performed by the provider.[*] Financial institutions must not share U.S. regulatory examination reports or information contained therein with either foreign regulators or foreign-based service providers without the express written approval of the appropriate U.S. regulatory authority.

Choice Of Law
Before entering into an agreement or contract with a foreign-based vendor or developer, an organization should carefully consider which country's law it wishes to control the relationship. Based on that review, organizations should include choice of law and jurisdictional covenants that provide for the resolution of disputes between the parties under the laws of a specific jurisdiction.

These provisions are necessary to maintain continuity of service, access to data, and protection of customer information. For these reasons, it can be particularly important when dealing with foreign service providers to specify exactly which country's laws will control the contractual relationship between the parties. Additionally, contract provisions may be subject to foreign-court interpretations of local laws. The laws of the foreign country may not recognize choice of law provisions and may differ from U.S. law regarding what they require of organizations

---

[1] 12 CFR part 364, Appendix B, ¶ III.D.2 - Banks and 12 CFR part 570, Appendix B, ¶ III (d)(2)- Thrifts.

[2] 12 CFR part 332 - Banks and 12 CFR part 573 - Thrifts.

[3] The term "U.S. regulatory authorities" means the FFIEC member agencies issuing this booklet.

[*] 12 USC 1867(c)(1)- Banks and 12 USC 1464(d)(7)- Thrifts. In addition, organizations should notify their primary regulatory authority of a service relationship with a foreign-based service provider in accordance with regulations and guidance issued by that regulator.

or how they protect bank customers. Thus, an organization's due diligence should include analysis of a country's local laws by legal counsel competent in assessing the enforceability of all aspects of a contract.

MONITORING AND OVERSIGHT

Monitoring foreign entities requires the same steps as monitoring domestic servicers and vendors in addition to the recommendations presented within this appendix. When organizations establish a servicing arrangement with a foreign-based service provider, management should monitor both the entity and the conditions within the foreign country.

The organization should determine that the foreign-based service provider maintains adequate physical and data security controls, transaction procedures, business resumption and IT contingency arrangements (including periodic testing), insurance coverage, and compliance with applicable laws and regulations. Further, where indicated by the organization's security risk assessment, the organization must monitor its foreign-based service providers to confirm that they have satisfied security obligations imposed in the contract to comply with Section 501(b) of GLBA.

Organizations also should monitor economic and governmental conditions within the foreign country to determine whether changes are likely to affect the ability of the service provider to perform under the arrangement.

REGULATORY AGENCY ACCESS TO INFORMATION

U.S. regulatory authorities must have the ability to examine the services performed by an organization's third-party service provider regardless of whether it is foreign or domestically based. Organizations must maintain, in the files of a U.S. office, appropriate English language documentation to support all arrangements with service providers. Appropriate documentation typically includes a copy of the contract establishing the arrangement, supporting legal opinions, due diligence reports, audits, financial statements, performance reports, and other critical information.[1] In addition, the organization should have an appropriate contingency plan to ensure continued access to critical information, to maintain service continuity, and the resumption of business functions in the event of unexpected disruptions or restrictions in service resulting from transaction, financial, or country risk developments.

---

[1] In instances where the financial institution's foreign branches have outsourced local operations or services cross-border to third-party service providers domiciled in another foreign country, copies of such records can be maintained at the foreign branch office, but must also be available in the U.S.

EXAMINATION CONSIDERATIONS

U.S. regulatory authorities may examine the services performed for an organization under an outsourcing arrangement with a foreign-based service provider. Likewise, in the case of a foreign-regulated entity, U.S. regulatory authorities may be able to obtain information through the appropriate supervisory agency in the service provider's home country.

With respect to the outsourcing organization in such arrangements, U.S. regulatory authorities will focus reviews on the adequacy of an organization's due diligence efforts, its risk assessments, and the steps taken to manage those risks including the effect of the arrangement upon the organization's compliance with applicable laws and its access to critical information. Regulatory reviews will assess the organization's contract provisions and its ongoing monitoring or oversight program, including any internal and external audits arranged by the foreign-based service provider or the organization.

An organization's use of a foreign-based third-party service provider (and the location of critical data and processes outside of U.S. territory) must not compromise the ability of U.S. regulatory authorities to effectively examine the organization. Thus, organizations should not establish servicing arrangements with entities where local laws or regulations would interfere with U.S. regulatory agencies' full and complete access to data or other relevant information. Any analysis of foreign laws obtained from counsel should include a discussion regarding regulatory access to information for supervisory purposes.

Index

Printed in the United States
by Baker & Taylor Publisher Services